Women in Engineering and Science

Series Editor

Jill S. Tietjen, Greenwood Village, USA

The Springer Women in Engineering and Science series highlights women's accomplishments in these critical fields. The foundational volume in the series provides a broad overview of women's multi-faceted contributions to engineering over the last century. Each subsequent volume is dedicated to illuminating women's research and achievements in key, targeted areas of contemporary engineering and science endeavors. The goal for the series is to raise awareness of the pivotal work women are undertaking in areas of keen importance to our global community.

Jenna P. Carpenter
Editor

Women in Academia

Springer

Editor
Jenna P. Carpenter
School of Engineering
Campbell University
Buies Creek, NC, USA

ISSN 2509-6427 ISSN 2509-6435 (electronic)
Women in Engineering and Science
ISBN 978-3-031-97937-8 ISBN 978-3-031-97938-5 (eBook)
https://doi.org/10.1007/978-3-031-97938-5

This Springer imprint is published by the registered company Springer Nature Switzerland AG
The registered company address is: Gewerbestrasse 11, 6330 Cham, Switzerland

Dedicated to all of the academic women in STEM who have come before us, for trailblazing our paths, and to those who have supported their success. May we build on their legacy to create a better future for those girls and women who follow us.

Foreword

Progress of the participation and leadership roles of women in engineering and computer science is not linear, contrary to some beliefs that time will fix this. In my 50 years of involvement in these fields and in advocacy roles to increase women's participation, not only in numbers but also on the impact women can have on the culture and environment, I have seen positive outcomes in some decades and regress at other times (Frize, 2009, 2019).

In our current historical period, women have made great gains, and it is hard to imagine that their position could regress. If we are to learn from history, however, we ought not to ignore the fact that, along with the great gains that women have made in academia and science, a critical mass of women has still not been seen in certain traditional fields of science and engineering, and women are conspicuously absent from several new scientific fields such as nanotechnology and robotics. The so-called pipeline model, which assumed that if women filled the lower positions, they would eventually share the top positions, has been acknowledged as a failure (Schiebinger, 1999, Frize, 2009).

There are at least two ways to look at the history of women and science. The first way is to look at the issue from an external perspective and explore the exclusion of women from the practice of science. The second way is to look at the issue from an internal perspective and explore the practice and content of science itself with an eye to uncovering possible biases (Frize, 2009).

In my view, progress and regress depend on several factors: the policies and will for change at the top, whether this is at the level of government, universities, popular culture, parents' and teachers' attitude, and support in encouraging girls and young women to consider and explore nontraditional careers like engineering and computer science. Professional and scientific associations can play a key role in encouraging women to take on leadership roles and advocate for change where it is needed. It is also important for men to participate in the efforts to reach the goals of equity, diversity, and inclusion (EDI).

The book *Women in Academia* provides an important contribution to the examination and understanding of the historical, current, and future opportunities and challenges to increase the participation of women in academic roles in these fields.

The book documents important developments such as the creation of associations like the Society of Women Engineers (SWE) and Women in Engineering ProActive Network (WEPAN) and programs like ADVANCE by the National Science Foundation (NSF). Women leading scientific and professional associations and interviews with women members of the National Academy of Engineering provide insights on women's important potential contributions to the culture and development of these fields.

Reaching equity would enrich the diversity of experiences and thus provide new approaches to education and development of technological solutions (Belenky et al., 1986; Shepherd, 1993). But we are currently facing a huge challenge with the dismantling of DEI programs in the United States in 2025. Reading this book and using it as a base to develop innovative approaches would help to continue major efforts to recruit and retain women in academic programs in engineering and computer science. These efforts may need to be relabeled under new names and must continue, otherwise regress will quickly take over. Everyone involved in efforts to increase the participation of women and their role in engineering and computer science will benefit greatly from reading this book and thinking of ways to build on past efforts and thus ensure progress.

Monique Frize
Faculty of Engineering and Design, Carleton University
Ottawa, Canada

References

Belenky, M. F., Clinchy, B. M., Goldberger, N. R., & Tarule, J. M. (1986). *Women's ways of knowing: The development of self, voice, and mind.* Basic Books.

Frize, M. (2009). *The Bold and the Brave: A history of women in science and engineering*. University of Ottawa Press.

Frize, M. (2019). *A Woman in Engineering: Memoirs of a Trailblazer.* University of Ottawa Press.

Schiebinger, L. (1999). *Has Feminism Changed Science?* Cambridge MA: Harvard University Press.

Shepherd, L. J., (1993). *Lifting the Veil: The Feminine Face of Science*, Shambhala Publications. ISBN 978-1570626821.

Preface

Several years ago, I stopped by a booth at the Society of Women Engineers annual conference to talk to Jill Tietjen. She was editing a series of books on women in engineering and science. I had previously worked with Jill to write several articles for a series on the history of women in the engineering honor society Tau Beta Pi. I asked Jill where the book on *academic* women in engineering and science was. She replied that there was not yet a book on that topic, but that it was a great idea. Since I am writing this preface and you are now reading this book, you can guess the rest of the story.

It was a daunting task to narrow down a broad topic like academic women in engineering and science, so I started my job as editor by interviewing a number of notable women in engineering and computing around the country to gauge interest in the volume, glean ideas on possible chapter topics, and recruit authors. Looking back, I wish I had asked permission to record the interviews! They were fascinating. The conversations with these women pioneers and leaders definitely shaped this book, and I was fortunate to engage many of them as authors. Given the impact that these authors have had, you will see many of them referenced in multiple other chapters of this volume.

While we have made some progress in increasing the number of women in engineering and computing in recent decades, it has been slow and remains far short of the goal of attracting all of the engineering talent that resides in the women of our nation. We are facing a well-documented demographic decline in the United States. Add to that the retirement of Baby Boomers, and it should not be a surprise that companies who employ engineers and computing professionals are already finding it difficult to hire the talent they need. While automation and AI may replace some jobs, workforce needs in engineering and computing are forecast to grow, not decline. Engineering skill sets like creative problem-solving are predicted to remain in high demand even in an AI-driven workforce. Our nation cannot meet its engineering workforce needs (and thereby maintain our socioeconomic status and national security, among other things) unless we figure out how to attract and retain women in engineering and computing in large numbers. To do that, we must heed the lessons and advice contained in the pages of this book. As it turns out, our collective future depends on it.

Buies Creek, NC, USA Jenna P. Carpenter

Acknowledgements

Thanks to Jill Tietjen for giving us the opportunity to highlight the history and success of academic women in engineering and science as well as their struggles and the challenges that remain. Thanks to each of the authors and the interviewees for their willingness to share their expertise and stories.

Introduction

Abstract
This introduction provides a brief overview of women in academic engineering in the United States and a summary of the chapters in this volume that explore the topic in more depth and from various perspectives.

This volume appears at a critical time for women, members of other underrepresented groups, and those who advocate for broadening participation in the US science and engineering workforce. In the turbulent early months of 2025, institutions and programs that have supported increased equity and diversity in society in general and science and engineering in particular are being attacked, defunded, and redefined. Decades of progress are under threat. At such a time, it is especially important to highlight the leadership and value that women have brought to engineering and academe. The contributors to this volume document through their scholarship and personal reflection the contributions of women to the engineering profession, engineering education, and engineering culture and provide insights for the future.

Until the middle of the eighteenth century, engineers learned their craft on the job through apprenticeship and hands-on experience. In the early years of the United States, many civil engineers learned their trade on the construction of the Erie Canal and other infrastructure projects. The first school in the United States designed to train engineers was the military academy at West Point, established by the federal government in 1802. As engineering professionalized later in the nineteenth century, formal education for engineers first became more common, then required. The land grant universities were chartered by Congress during the Civil War specifically to train young men in the agricultural and mechanical arts and have evolved into some of the largest engineering schools in the country (Reynolds & Tietjen, 2001; Bix, 2004).

The first record of an engineering degree awarded to a woman in the United States was from the University of California at Berkeley in 1876 when Elizabeth Bragg earned a degree in civil engineering, but there is no documentation of her actually practicing the profession. A few women earned degrees and worked as engineers in the late nineteenth and early twentieth centuries, but they remained few and far between. Margaret Ingels' delightful 1952 essay "Petticoats and Slide Rules" describes a parade of women engineering graduates in the first half of the twentieth century, but the numbers of women earning degrees in engineering each year remained so small that they could be enumerated by single digits (Ingels, 1952).

World War II opened many doors for women. Not only were women called on to work on the assembly lines like Rosie the Riveter, but they were also recruited into engineering schools to learn drafting and other technical skills. Aircraft manufacturer Curtiss-Wright Corporation established training programs for women at seven engineering colleges, recruiting women who had little familiarity with engineering as well as those who had expressed interest in the field but had previously been discouraged from pursuing engineering as an unsuitable job for girls. The program provided an intensive introduction to aeronautical engineering and a job on completion of the program. However, both the training and the jobs were designed as temporary support for the war effort and did not necessarily provide a path to an engineering career after the war. Some of the women trained in this and other wartime programs did go on to find careers in technical fields, but after the war ended, women were encouraged to leave the workforce to make room for returning soldiers (Bix, 2004, 2013).

While the number of women studying engineering remained small through the middle of the twentieth century, the number of women engineering faculty members was even smaller. Ellen Swallow Richards taught chemistry at MIT from the 1870s until 1911, and her work in water quality and sanitation is often cited as providing a foundation for sanitary engineering, now environmental engineering, as well as home economics. Lillian Gilbreth taught engineering management and industrial engineering at Purdue from the 1930s and was the first woman elected to the National Academy of Engineering in 1965 (Trescott, 1990). Neither Richards nor Gilbreth held degrees in engineering, but their work clearly fit within current definitions of the field. Edith Clarke did earn a degree in electrical engineering from MIT in 1919 and became a professor at the University of Texas in 1945 after retiring from a career at General Electric (Brittain, 1985). Sheila Widnall and Mildred Dresselhaus joined the engineering faculty at MIT in the 1960s (Bix, 2013).

The US Departments of Labor and Education and engineering professional societies have collected data on the number of engineering degrees awarded to men and women since at least the middle of the twentieth century. Launch of the Sputnik satellite by the USSR in 1957 and the resulting space race increased attention to science and engineering education in the United States, but less than 1% of engineering degrees were awarded to women until the early 1970s. The civil rights and women's rights movements of the 1960s brought more women into higher education and the workforce overall, but few pursued engineering careers. The passage of Title IX of the Education Amendments of 1972 prohibiting sex discrimination in education programs receiving federal funds opened the doors of the military academies and other all-male colleges and universities, including many engineering programs, to women (Vetter, 1981).

Between 1972 and 1984, the number of engineering degrees awarded to women increased by a factor of 20, and the percent of engineering degrees earned by women reached 15%. By the turn of the twenty-first century, the percentage of engineering degrees earned by women topped 20%, and there it has stayed, more or less, for the last 25 years. The overall number of engineering degrees awarded has increased greatly during that time frame, but the participation of women as a percent of the overall engineering field has not changed significantly (Layne, 2009; ASEE, 2021).

In this volume, Bottomley et al. review these enrollment trends and present research that explores factors contributing to the persistent underrepresentation of women in engineering. Holloway et al. discuss the evolution of programs to support women engineering students as they entered US universities in increasing numbers in the 1970s and the establishment of the Women in Engineering ProActive Network. Originally intended to provide resources for those programs, the organization has expanded its focus to advocate for systemic change in the culture of engineering and engineering education to make it more inclusive, equitable, and diverse. Tietjen describes the history of women engineers in the United States in more depth, including the role of professional societies in defining and policing the boundaries of the profession and the establishment of the Society of Women Engineers to provide a welcoming environment. Stormes et al. focus on women in academic computing, trends in enrollment, and efforts of various organizations to increase women's participation, with varying results.

The National Science Foundation has had a mandate to promote the progress of science by ensuring an adequate scientific and technical workforce since its establishment in 1950. That mandate was explicitly expanded to include broadening participation of women, members of other underrepresented groups, and persons with disabilities by the passage of the Science and Engineering Equal Opportunity Act as part of the agency's reauthorization in 1980. The Act established the Committee on Equal Opportunities in Science and Engineering to advise the agency and produce biennial reports to Congress (NSF, 2024; Malcom, n.d.). After 20 years of programs supporting individual women in science and engineering, NSF took a radically different approach in 2001 with the establishment of the ADVANCE program focusing on the role of institutions in increasing the number and success of women faculty (Stewart et al., 2007). The chapter by Blevins and Blevins highlights the role of women leaders at the National Science Foundation, while De Aro et al. document the establishment and growth of, changes to, and lessons learned from 20 years of the ADVANCE program.

In 2001, the National Academies published an in-depth study of gender differences in the careers of doctoral scientists and engineers. Analyzing data from the last quarter of the twentieth century, the report is aptly titled *From Scarcity to Visibility*. The report concluded that during that time period, women made great progress toward equality in science and engineering but were still more likely than their male colleagues to have lower status and lower pay, with men more likely to be employed at research-intensive institutions, to hold tenured positions, and to be in leadership roles (National Research Council, 2001).

While the percent of tenured and tenure-track faculty in US engineering schools who are women is now approaching 20% and women make up 27% of assistant professors, less than 15% of faculty at the senior rank of professor are women (ASEE, 2023). Women began to move into leadership roles in engineering education in 1984 when Dr. Eleanor Baum was named Dean of Engineering at Pratt Institute. The number of women engineering deans and department heads grew slowly through the 1980s and 1990s, and more rapidly in the twenty-first century, with almost a quarter of US engineering programs led by women in 2021. Some of these women have gone on to serve as provosts, vice presidents, presidents, and chancellors at universities across the country (Layne, 2021).

The remaining chapters in this volume explore the differing experiences and perspectives of women engineers of varying ethnic backgrounds and nationalities in academic and professional leadership roles. Cross et al. provide a theoretical framework and personal reflections on the impact of intersectional identities on women leaders in engineering education. Rosser highlights the experiences of foreign-born and -educated women faculty in US engineering schools. Barabino reflects on her personal journey to the presidency of a unique engineering college and the various ways that identity shapes the career trajectories of women in academic engineering careers. Carpenter describes the inclusion of women in the National Academy of Engineering, while Watson and Adams explore differences in the history and culture of engineering professional societies and resulting lessons for women engineers in leadership roles.

Throughout this volume, occasional "callouts" highlight concepts, programs, and experiences that provide additional perspectives on women in academic engineering. Chubin's concluding chapter reflects on a lifetime of research, observations, and advocacy for broadening participation in science and engineering through the changing cultural and political landscape of the past half century. As a sociologist, Chubin points out that the concept of "merit" is malleable and subject to change, and he emphasizes the importance of change agents in calling out bias and shaping public understanding in order to continue to move forward. In today's contentious environment, it is more important than ever for advocates and allies to continue to communicate the contributions of women and the value of diversity to the engineering profession.

Virginia Tech
Blacksburg, VA, USA
playne@vt.edu

Peggy Layne

References

ASEE. (2021). *Profiles of Engineering and Engineering Technology*.

ASEE. (2023). *Engineering and Engineering Technology by the Numbers 2022*.

Bix, A. S. (2004). From 'engineeresses' to 'girl engineers' to 'good engineers': A history of women's U.S. Engineering Education. *NWSA Journal, 16*(1), 27–49.

Bix, A. S. (2013). *Girls Coming to Tech! A History of American Engineering Education for Women*. MIT Press.

Brittain, J. (1985). From computer to electrical engineer: The remarkable career of Edith Clarke. *IEEE Transactions on Education, E-28*(4), 184–189.

Ingels, M. (1952). *Petticoats and slide rules*. Midwest Engineer, August 2–4, 10–12, 15–16.

Layne, M. (2009). *Women in Engineering: Professional Life*. ASCE Press.
Layne, P. (2021). Women Engineering Leaders in Academe 2021. *SWE Magazine, 67*(4), 42–48.
Malcom, S. (n.d.) Reflecting on our Roots Part 1: How the NSF Committee on Equal Opportunities in Science and Engineering (CEOSE) Came to Be. NSF INCLUDES
National Research Council. (2001). *From Scarcity to Visibility: Gender Differences in the Careers of Doctoral Scientists and Engineers*. National Academy Press.
National Science Foundation. (2024). Committee on Equal Opportunities in Science and Engineering. Retrieved from https://new.nsf.gov/od/oia/ceose. Accessed 15 Apr 2024
Reynolds, B., & Tietjen, J. (2001). *Setting the Record Straight: The History and Evolution of Women's Professional Achievement in Engineering*. White Apple Press.
Stewart, A., Malley, J., & LaVaque-Manty, D. (Eds.). (2007). *Transforming Science and Engineering: Advancing Academic Women*. University of Michigan Press.
Trescott, M. M. (1990). Women in the intellectual development of engineering: A study in persistence and systems thought. In G. Kass-Simon & P. Farnes (Eds.), *Women of science: Righting the record* (pp. 147–165). Indiana University Press, 182–185.
Vetter, B. M. (1981). Women scientists and engineers: Trends in participation. *Science, 213*(4527), 1313–1321.

Contents

Contributors

Stephanie G. Adams University of Texas at Dallas, Richardson, TX, USA

Idalis Villanueva Alarcón University of Florida, Gainesville, FL, USA

Gilda A. Barabino Olin College, Needham, MA, USA

Lecia Barker University of Colorado Boulder, Boulder, CO, USA

Canan Bilen-Green North Dakota State University, Fargo, ND, USA

Emily G. Blevins Alexandria, VA, USA

Linda G. Blevins U.S. National Science Foundation, Alexandria, VA, USA

Laura J. Bottomley North Carolina State University, Raleigh, NC, USA

Rebecca Brent Education Designs, LLC, Chapel Hill, NC, USA

Dana M. Britton Rutgers University, New Brunswick, NJ, USA

Jenna P. Carpenter Campbell University, Buies Creek, NC, USA

Daryl E. Chubin Independent Consultant, Savannah, GA, USA

Kelly J. Cross Georgia Tech and Emory University, Atlanta, GA, USA

Christine Cunningham Museum of Science, Boston, MA, USA

Jane Z. Daniels Purdue University, West Lafayette, IN, USA

Jessie A. DeAro U.S. National Science Foundation, Alexandria, VA, USA

Roger Green North Dakota State University, Fargo, ND, USA

Beth M. Holloway Purdue University, West Lafayette, IN, USA

Peggy Layne Virginia Tech, Blacksburg, VA, USA

Kathleen J. Lehman University of South Carolina, Columbia, SC, USA

Elizabeth Litzler University of Washington, Seattle, WA, USA

Susan Staffin Metz Stevens Institute of Technology, Hoboken, NJ, USA

Carol B. Muller Stanford University, Stanford, CA, USA

Olga Pierrakos Wake Forest Engineering the National Science Foundation, Alexandria, VA, USA

Simil Raghavan National Academy of Engineering, Washington, DC, USA

Kathy Renzetti DiscoverE, Washington, DC, USA

Donna Riley University of New Mexico, Albuquerque, NM, USA

Sue V. Rosser San Francisco State University, San Francisco, CA, USA

Thea Sahr DiscoverE, Watertown, MA, USA

Keri Ann Sather-Wagstaff Framingham State University, Framingham, MA, USA

Denise R. Simmons University of Florida, Gainesville, FL, USA

Chrystal A. S. Smith University of Connecticut, Storrs, CT, USA

Kaitlyn N. Stormes University of Georgia, Athens, GA, USA

Jill S. Tietjen Technically Speaking, Inc., Greenwood, CO, USA

Karan L. Watson Texas A&M University, College Station, TX, USA

Abbreviations

A&A	Advocates and Allies
AAAS	American Association for the Advancement of Science
ABET	(formerly) Accrediting Board for Engineering and Technology
ACE	American Council on Education
ACM	Association for Computing Machinery
ACM-W	ACM Committee on Women in Computing
ADVANCE	(NSF) Increasing the Participation and Advancement of Women in Academic Science and Engineering Careers Program
AIChE	American Institute of Chemical Engineers
AIEE	American Institute of Electrical Engineers
AIICE	Alliance for Identity-Inclusive Computing Education
AIME	American Institute of Mining Engineers
ARC Network	NSF ADVANCE Resource and Coordination Network
ARC VVS	ARC Visiting Virtual Scholar
ASCE	American Society of Civil Engineers
ASEE	American Society for Engineering Education
ASME	American Society for Mechanical Engineering
B.C.E.	Before Common Era
BIPOC	Black, Indigenous, and People of Color
BME	Biomedical Engineering
BMS	(NSF) Biological and Medical Sciences Division
C.E.	Common Era
CARES Act	Coronavirus Aid, Relief, and Economic Security Act
CCNY	City College of New York
CDI	(NSF) Cyber-enabled Discovery and Innovation Program
CEO	Chief executive officer
CITRIS	(University of California, Berkeley) Center for Information Technology Research at the Interest of Society
CoNECD	Collaborative Network for Engineering and Computing Diversity
Co-PI	Co-Principal Investigator
CRA	Computing Research Association

CRA-E	(CRA) Committee on Education
CRA-W	(CRA) Committee on the Status of Women in Computing
CRA-WP	(CRA) Committee on Widening Participation
CS	Computer science
CUNY	City University of New York
CV	Curriculum vita
CW Cadettes	Curtis-Wright Cadettes
DAD	(NSF) Deputy Assistant Director
DEI	Diversity, Equity, and Inclusion
DEIA	Diversity, Equity, Inclusion, and Access
DREU	(NSF) Distributed Research Experiences for Undergraduates
ECR	(NSF) Education Core Research
EDI	Equity, Diversity, and Inclusion
EDT	Emergency Defense Training
EER	Engineering Education Research
EPSCoR	(NSF) Established Program to Stimulate Competitive Research
ESMWT	Engineering, Science, and Management War Training
EXITE	Exploring Interest in Technology and Engineering
FIPSE	Fund for Improvement of Postsecondary Education
FIU	Florida International University
GHC	Grace Hopper Celebration of Women in Computing
GIs	Government or General Issue (Army-enlisted soldier in World War II)
GSOE	Grove School of Engineering
HBCU	Historically Black Colleges and Universities
iAAMCS	Institute for African American Mentoring in Computing Sciences
IBM	International Business Machines
IBM/HGST	IBM/Hitachi Global Storage Technologies
ICES	(NSF) Interface Between Computer Science and Economics and Social Science Program
I-CORPS	(NSF) Innovation Corps Program
IDEALS	Inclusion, Diversity, Equity, Accessibility, and Leadership Skills
IEEE	Institute of Electrical and Electronics Engineers
IHE	Institution of Higher Education
IIP	(NSF) Industrial Innovation and Partnerships Division
INCLUDES	(NSF) Inclusion across the Nation of Communities of Learners of Underrepresented Discoverers in Engineering and Science
INSPIRE	Purdue's Research Institute for Pre-College Engineering
IoT	Internet of Things
IT	(NSF ADVANCE) Institutional Transformation
JPL	Jet Propulsion Lab
JSU	Jackson State University
KSU	Kansas State University
LGBTQ	Lesbian, Gay, Bisexual, Transgender, Queer/Questioning

LGBTQIA+	Lesbian, Gay, Bisexual, Transgender, Queer/Questioning, Intersex, Agender/Asexual, etc.
MIT	Massachusetts Institute of Technology
MPE	(NSF) Mathematical, Physical, and Engineering Sciences Division
NAE	National Academy of Engineering
NAM	National Academy of Medicine
NAMEPA	National Association of Multicultural Engineering Program Advocates
NASA	National Aeronautics and Space Administration
NASEM	National Academies of Science, Engineering, and Mathematics
NAVAIR	Naval Air Systems Command
NCES	National Center for Education Statistics
NCWIT	National Center for Women & Information Technology
NDSU	North Dakota State University
NEC	Nippon Electric Company
NRC	National Research Council
NRI	National Robotics Initiative
NSBE	National Society of Black Engineers
NSF	National Science Foundation
NSPE	National Society of Professional Engineers
OIA	(NSF) Office of Integrative Activities
OISE	(NSF) Office of International Science and Engineering
OWI	Office of War Information
PAESMEM	Presidential Award for Excellence in Science, Mathematics and Engineering Mentoring
PAID	Partnerships for Adaptation, Implementation, and Dissemination
PhD	Doctor of Philosophy
PI	Principal investigator
PIC	ASEE Professional Interest Council
PRISMA	Preferred Reporting Items for Systematic Reviews and Meta-Analysis
PWI	Predominantly White Institution
RCA	Radio Corporation of America
REU	Research Experience for Undergraduates
SaTC	(NSF) Secure and Trustworthy Cyberspace Program
SBE	(NSF) Social, Behavioral, and Economic Sciences Directorate
SCH	(NSF) Smart and Connected Health Program
SEA Change	(AAAS) STEMM Equity Achievement Change
SHPE	Society of Hispanic Professional Engineers
SISC	Single Institution Systemic Change
SME	Society for Mining, Metallurgy and Exploration
SoC	(NSF) Social Computational Systems Program
SPE	(NSF) Scientific Personnel and Education Division
STEM	Science, Technology, Engineering, and Mathematics
STEMM	Science, Technology, Engineering, Mathematics, and Medicine

STRATEGIC	Strategies for Effecting Gender Equity and Institutional Change
STRIDE	Strategies and Tactics for Recruiting to Improve Diversity and Excellence
SU	Seattle University
SUNY	State University of New York
SWE	Society for Women Engineers
TA	Teaching assistant
TAMU	Texas A&M University
TECAID	Transforming Engineering Culture to Advance Inclusion and Diversity
TIP	(NSF) Technology, Innovation, and Partnerships Directorate
UCLA	University of California at Los Angeles
UMD	University of Maryland
UNH	University of New Hampshire
UR2PhD	(CRA) Undergraduate Research to PhD Program
URM	Underrepresented minority
USA	United States of America
USB/UST	US-born/US-trained
USC	University of Southern California
USSR	Union of Soviet Socialist Republics
UTRGV	University of Texas Rio Grande Valley
UU	University of Utah
WAAIME	Women's Auxiliary to the AIME
WECE	Women's Experiences in College Engineering
WEPAN	Women in Engineering ProActive Network
WiE	Women in Engineering
WIED	(ASEE) Women in Engineering Division
WISE	(University of Washington) Women in Science and Engineering
WKC	WEPAN Knowledge Center
WPI	Worcester Polytechnic Institute

Chapter 1
Women in Engineering Education

Laura J. Bottomley, Simil Raghavan, Kathy Renzetti, Christine Cunningham, and Thea Sahr

1.1 Introduction

Although we do not know the names, genders, or educational backgrounds of the earliest engineers, we do know that around 6000 B.C.E., irrigation systems were invented in the region that would become China. By 2900 B.C.E., Egyptians were building pyramids, and advancements in road construction, building design, and agriculture were being made.

In the Middle Ages (circa 1325 C.E.), there were two recognized categories of engineering—military and civil—with the term engineer being defined as "a constructor of military engines." Leonardo da Vinci offered his services as an engineer to the Senate of Venice when they were at war with the Ottoman Empire. The French established the first engineering school in 1747 (Furter, 1980), paving the way for the Industrial Revolution.

L. J. Bottomley (✉)
North Carolina State University, Raleigh, NC, USA

S. Raghavan
National Academy of Engineering, Washington, DC, USA

K. Renzetti
DiscoverE, Washington, DC, USA
e-mail: Kathy@DiscoverE.org

C. Cunningham
Museum of Science, Boston, MA, USA
e-mail: ccunningham@mos.org

T. Sahr
DiscoverE, Watertown, MA, USA

J. P. Carpenter (ed.), *Women in Academia*, Women in Engineering and Science,
https://doi.org/10.1007/978-3-031-97938-5_1

By the eighteenth century, engineering had established itself as a profession, and the term engineer came to mean using math, science, and other knowledge to solve problems. Fields like mechanical engineering, electrical engineering, and chemical engineering were added in the nineteenth and twentieth centuries. (For example, the first electrical engineering department and degree program in the world began in Darmstadt, Germany, in 1882. A year later, Cornell and MIT initiated the second program as well as a long-standing rivalry about whose program came first.) Echoing how society was structured, the problems engineers solved were those identified by men as necessary for the military, society, and industry—women were decidedly not welcome.

This historical perspective illustrates some important points. Namely, engineering began in contexts that were masculine. With a few exceptions, women were prohibited from studying engineering throughout the nineteenth century. Often, women inventors received their training in other fields of mathematics or science. In the early years of the twentieth century, women in engineering programs were viewed as a curiosity, but the advent of World War II and major shortages of engineering talent led to new training programs within the USA to engage women in engineering and technical careers. Although wartime efforts to engage women were effective, these women were still viewed as temporary substitutes for men. Women continued to face resistance following the war as efforts were made to reestablish prewar social norms (Barker, 2003).

As women gained admittance to engineering programs during the 1940s, many began organizing local meetings or networking groups within their universities. In particular, graduate students at Drexel, Cooper Union, and City College of New York formed student groups that later became the Society of Women Engineers (SWE). Dr. Beatrice Hicks became the first elected president of SWE in 1950. The organization grew slowly through the 1950s but doubled its membership in the 1960s as the government's increasing commitment to technological development, along with societal pressure, encouraged more schools to open their doors to women.

Slowly, as evidenced by the timeline presented here, women inserted themselves into engineering and engineering education. The advent of women did not necessarily lead to change. Instead, through careful, concerted effort, women and their collaborators have worked deliberately to modify systems and the profession. In the twenty-first century, what it means to be an engineer is evolving. The role of ethics, needs of humanity in the face of climate change, and social justice issues around who benefits and who does not from our engineering solutions are increasingly being considered by the profession. This expanded view requires the diversification of the engineering mindset and the recognition that tapping a range of talents and perspectives results in more innovative solutions. It requires a new definition of engineering. It requires women, all people of color, and people from all socioeconomic levels be encouraged and welcomed into the profession (Fig. 1.1).

A Timeline of Firsts for Women in Engineering Education in the U.S.

The story of women in engineering education mirrors that of women in engineering in general. We have a few examples of women who worked as engineers, many without specific engineering degrees, but they are notable by their scarcity. Precisely identifying who was the "first" in engineering history can be complex, given that engineering terminology was not always used to describe their position and because women were often written out of history. A particular challenge is also identifying non-white women and the role that they played in history. Their contributions were more frequently marginalized, and sometimes they hid their identities in order to be taken more seriously. With that in mind, here are some probable firsts for women in the United States.

1876: First engineering bachelor's degree: Elizabeth Bragg (Civil engineering, University of California, Berkeley)

1890: First engineering instructor: Ellen Swallow Richards (Industrial/safety engineering, MIT)

1917 First engineering professor: Helen Dodson Prince (Civil engineering, Smith College)

1931: First PhD in engineering : Yun-Hao "Ruth" Feng (Chemical engineering, The Ohio State University)

1936: First Tau Beta Pi Women's Badge recipient (women were not eligible for full membership until 1969): Katharine Cleveland Harelson (Civil Engineering, University of Kentucky, graduated in 1924)

1946: First bachelor's degree to a Black woman (Civil engineering): Hattie Peterson (Civil engineering, Howard University)

Fig. 1.1 A timeline of firsts for women in engineering education in the USA

1.2 What Has Had an Impact?

There is an abundance of research on why women choose *not* to stay in engineering majors in higher education or the reasons women avoid or leave engineering pathways. A research paper published by DiscoverE sheds light on the women who *persisted to become engineers*. The paper summarizes what the engineering education literature says about (Ayre et al., 2013) why young women *choose* to pursue engineering in college and (Baker & Leary, 1995) what factors affect whether young women *persist* in engineering.

When possible, the research also explored literature about whether these factors are different for women of different races, ethnicities, sexual orientations, gender identities, and socioeconomic classes. This attention to intersectionality is incredibly important, because women are not a monolith! Individual experiences vary by ethnicity, gender identity and sexuality, family background, and more.

What follows are seven key factors that DiscoverE identified in its literature review that impacted women and girls' decisions to persist in an engineering pathway. The report is aptly titled *Despite the Odds* (DiscoverE, 2019).

1.2.1 Key Factors

As discussed in the brief historical view above, engineering started out as majority male. However, the field has grown and changed to become much more human centered and much more multidisciplinary. As harder and harder problems are addressed, women are more and more attracted to the field. If this is true, then why are undergraduate programs still only about 20% female? The reasons are complex.

One of the ways that the decisions young people make about career pathways have been modeled is that of a pipeline. Recent thinking, however, has realized that this model has constrained the way we examine both root causes and potential solutions to the diversification of engineering. A different model is that of a garden ecosystem, which invites consideration of factors like whether young people are exposed to real-world engineering or are encouraged to consider engineering as a potential career. It may even consider whether young people are taught factual information about what engineering is or what engineers do (Bottomley, 2022).

Once a young person is attracted to the field of engineering, there are multiple sources of influence that may prevent them from studying engineering, taking an engineering job, and persisting through each. It may be that an initial attraction to engineering shifts to something like theoretical physics once a student learns more about the world. But lack of confidence, support, or even people that look like them can also cause students to leave a field where they would be happy and successful if they had persisted.

Decades of research tell us what kinds of fertilizer an engineering ecosystem needs. Young women who choose and persist in engineering:

1. *Have positive attitudes about engineering*: these positive attitudes include favorable views of engineers and the impact of engineering in the world.
2. *See value in the discipline of engineering*: connecting engineering to the solution of important problems in the world can lead students to believe that engineering will enable them to make a positive impact on society and help people.
3. *Demonstrate engineering-related self-efficacy*: students who persist believe that they have the skills and knowledge to do engineering and that they are capable of contributing to the work of engineers. Self-efficacy is a tool that allows persistence through failure.
4. *Establish an engineering identity*: being able to relate to and see themselves as an engineer (solving problems, belonging to professional groups, and embracing the engineering mindset of creativity, innovation, ethics, and resilience to failure) contribute to embracing the idea that they are, or will someday become, engineers.

5. *Have a strong support network*: having the support of friends, family, peers, and/or role models is essential to all kinds of success. Students who are first generation in college can be vulnerable in this area, as their family may not understand the kinds of support that are needed. In addition, women who are entering a field that their family and friends view as predominantly masculine may need the support of mentors and role models to normalize their experiences.
6. *Draw upon social and cultural capital*: humans draw strength from personal or cultural experiences of struggle to overcome obstacles. Social and cultural capitals include norms and values that align as well as networks of relationships where support and connections can be found. Having a network to help find an internship is one example of social capital. Cultural capital can be placing value on education. Combined, these things can strongly influence students' expectations about their educational and work experiences.
7. *Feel a sense of belonging*: feeling as if one has found a place to belong is important for everyone. For women in engineering, finding a place where one feels valued in the community of engineers/engineer students is especially important. This can combat microaggressions that women experience, like being told, "You don't look like an engineer." Understanding that the engineering community has a place for all types of people is important for both community members and for those outside of the immediate community.

Research supports each of the above attributes of women who choose and persist in engineering. Understanding this research can lead to the development of programs and actions that will increase the percentage of women in engineering.

1.2.2 Interest and Attitudes

Researchers have demonstrated a connection between early interest in STEM (as early as elementary school) and subsequent choices to major in STEM in college. Maltese and Tai (2011) found that, as early as eighth grade, students who indicate an interest in STEM are highly likely to earn an undergraduate degree in a STEM field. Sadler et al. (2012) found that the key factor predicting STEM career interest at the end of high school was interest at the start of high school. Maltese (2008) examined how high school course taking influenced student college majors and degrees. He found that the most important factor related to STEM persistence was a positive attitude.

What this means for programming is that engaging interest is key. Engineering programming needs to be available and engaging to all students at a young age. Girls and young women need to see examples of engineering in their everyday experiences and as a part of things that they value. If they like cooking, then connections to chemical engineering are natural. If they like streaming content, then computer science and electrical engineering should be connected. These connections and experiences need to be not only special, extracurricular activities, but also a part of schooling and media so that engineering is seen as an option regardless of any particular privilege that a young woman might possess.

1.2.3 Value

Some research has focused on the relationship between attainment value, utility value, and STEM persistence. Attainment value is the importance students attach to a choice as it relates to their conception of identity and ideals or their competence in a given domain (Wigfield, 1994). STEM utility value is the importance students assign to a choice because it is relevant or useful to a student's pursuit of a current or future goal. Andersen and Ward (2014) analyzed the HSLS:09 data to look at the STEM persistence plans for ninth-grade, high-ability students. They found that ninth-grade, high-ability students who have a higher attainment value for science are more likely to plan to persist in STEM.

Numerous evaluations have found that out-of-schooltime programs like *Future City*, *Dream Big*, and *Invent It. Build It.* can help students make connections between their own values (e.g., wanting to help people) and the work done by engineers (Paulsen et al., 2021). However, more empirical research is needed in this area to definitively link values with the choice to pursue engineering.

Access to out-of-school programming is extremely important for students in the K-12 space, and programs like *Future City*, which has very low associated cost, are important for states to support, particularly in rural areas. But the need for value does not stop at the end of high school.

Colleges and universities need to ensure that students see the value proposition for engineering throughout the undergraduate experience. In traditional engineering programs, the first year consists of many courses that essentially fill students' toolboxes in preparation for second-year engineering courses. And even second- and third-year courses can lean toward the theoretical. Ensuring that students see connections to the kinds of problems around them is key to both recruiting AND retaining diverse engineering students. Programs making these links can be both curricular and cocurricular but need to extend across the entire undergraduate experience.

1.2.4 Self-Efficacy

Buontempo et al. (2017) conducted a study to see which factors influenced engineering identity ("I see myself as an engineering person"). The study found that self-efficacy was one of several significant predictors of engineering identity. Young and Young (2018) conducted a study that focused on African-American females and the factors that contributed to whether they enrolled in the advanced courses necessary for pursuing engineering degrees in college. The study found that self-efficacy was a positive predictor of enrollment in advanced high school science courses for

African-American females. The relationship between self-efficacy and choice to study engineering in college is an area that is ripe for more research.

There is strong evidence for the relationship between self-efficacy and persistence in engineering at the college level. For example, Marra et al. (2012) found that high self-efficacy was related to women students' plans to persist in engineering. Likewise, Cole and Espinoza (2011) found that women who reported high self-efficacy were more likely to report having such goals as pursuing a STEM career or attending graduate school. A study conducted by Naphan (2016) exploring young women's persistence in engineering programs also found that self-efficacy was the most important factor in predicting women's persistence in the program.

Some studies point to the importance of different types of self-efficacy for retention. Raelin et al. (2014) found that academic self-efficacy was significantly related to retention in college engineering programs. Work self-efficacy (tied to students' participation in co-op and internship programs) was also an important factor in retention. The study found that higher retention was positively correlated with the number of co-ops taken by students. Related to work self-efficacy is the concept of role confidence. Cech et al. (2011) found that if women develop more confidence about their abilities to be successful professionals and express less ambiguity about their fit or comfort within the discipline, then they remain in engineering at higher rates than expected.

Support for internships and co-ops is one way that undergraduates can develop engineering self-efficacy. Programming that illustrates engineering jobs and how they are related to coursework and participation in organizations like *Engineers Without Borders*, *Girls Engineering Change*, or *Alternative Service Breaks* can also have a strong influence on engineering self-efficacy as students realize the impact their work can have on others.

Academic self-efficacy can be supported by programs that normalize failure and recovery. This can be done in the classroom with pedagogical techniques like low-stakes assignments and opportunities to redo assignments. Co-curricular programming that emphasizes growth mindset and normalizes failure (perhaps by successful engineers talking about their own experiences with failure) can also have a strong impact.

1.2.5 STEM Identity

As explained by Godwin and Lee (2017), engineering identity is considered a crucial indicator for educational and professional outcomes that are important in engineering education. Recent research provides evidence that young women who choose engineering pathways are likely to start seeing themselves as engineers as early as middle or high school. Godwin found that identifying with math and physics upon entrance to college predicted engineering choice (for both men and women). Godwin argues that by "fostering the development of these subject-related identities prior to college enrollment and early in students' college careers, more students may be recruited and retained in engineering (p. 332)." Young and Young (2018) found that science identity had the highest correlation with advanced science coursework and that science identity was a positive predictor of enrollment in advanced high school science courses for African-American females.

The relationship between STEM identity and persistence is an area that needs more exploration at the college level. There is some evidence of a connection. For example, Gregory (2015) found six keys to the persistence of a group of female African-American engineering students: "(a) active involvement with the Black community on campus; (b) a strong desire to give back and inspire the next generation of engineers; (c) faith, family, and community; (d) a firm identity/strong sense of self; (e) being proud/passionate/committed to being an engineer; and (f) being advocates for themselves (p. v)." More research is needed to explore this topic further.

One way to impact STEM identity is to demystify topics like physics and math at an early age. The trope of "when will I ever use this" should not go unanswered. This work needs to take place both in the classroom and outside of the classroom.

1.2.6 Support Networks

A number of studies describe the importance of parental influence on high schoolers' interests and plans for college. Sahin et al. (2018) found that the students who were most likely to choose a STEM major in college were also most likely to be students with high parent encouragement. Kim et al. (2017) found that parents' expectations, their level of participation in their children's schools, and their own STEM engagement were all positively related to students' postsecondary enrollment. The authors contend that parents play a critical role in their children's academic and postsecondary outcomes. Based on their own research, Rozek et al. (2017) argue that parents are an important resource that can be engaged to increase motivation and promote important student outcomes.

A study by Dennehy and Dasgupta (2017) found that female engineering students who were provided with female mentors early in their college careers formed connections with the mentors, cementing, in effect, the young women's "sense of belonging." Fully 100% of the students provided with female mentors stayed in their engineering courses, while 89% of female students without mentors and only 82% of students with male mentors remained. The study found that female peer mentoring during the transition to college was effective at increasing belonging, confidence, motivation, and ultimately retention of women in engineering.

Some researchers have found that a student's mentors or role models do not necessarily have to be faculty (Griffith, 2010; Lourens, 2014). Griffith (2010) found that the existence of a higher percentage of female STEM graduate students positively impacted the persistence of female undergraduate students in STEM majors.

According to Ong (2011), family and community support is the most salient factor that women of color identify as fostering their completion of a STEM degree.

Trenor et al. (2008) looked at persistence and retention. They found that within an already-diverse environment, where racial barriers have been essentially mitigated, students of all ethnicities experience strong social support and weakened barriers. They also found that parents and role models played a significant role in female students' selection of and persistence in engineering programs.

Family and community support are important at all levels, so programming that helps educate family on the types of support that are most valuable can have a significant effect. Programs for parents of high school students that teach the true nature of engineering, as opposed to misconceptions about the field, are very important. Similarly, programs that engage the parents of first-generation college students and suggest strategies for supporting their children through difficulties, as well as educate them on the kinds of difficulties their students are likely to face, can be very valuable in retaining engineering students of all types.

1.2.7 Social Capital

There is emerging evidence from small, qualitative studies that there is a relationship between social capital and persistence. Agbenyega (2018) explored the persistence of Latina women from their undergraduate experiences, through attaining their engineering degree, to their early workplace experiences. She found that Latinas attending institutions in the USA drew upon multiple forms of social and cultural capital to overcome obstacles related to being female and non-white in a male-dominated field. She also found that women in the study drew "strength from personal or inherited experience of struggle along their trajectories through engineering. They face(d) challenges with an awareness that obstacles are part of any journey and developed both an ability to transform obstacles into inspiration for working harder and an understanding that overcoming them is crucial both to repaying the debt of sacrifice of those who came before and paving the way for those who will come after (pp. ii–iii)." This is an area that needs more empirical research. For students who enter college with less social capital, Women and Minority Engineering programs can help create support networks and professional networks that can make up the difference.

The timing of this chapter could not be more critical as recent events in the USA are undermining equitable and inclusive access for underserved communities, including the recent U.S. Supreme Court ruling (June 29, 2023) that effectively eliminated the use of affirmative action in college admissions. In addition, numerous states are taking similar measures to eliminate or reduce access to funding for organizations and engineering education programs serving underrepresented communities.

According to Joelle Emerson, CEO of Paradigm, and as reported by Erica Pandey in her article *Anti-DEI movement expands in politics, business and academics* for axios.com, we are seeing a new trend. "Those critiquing DEI aren't just the extreme, right wing anti-progress activists, like the group who challenged affirmative action," said Emerson. "They're also liberal leaning people who are likely aligned with DEI in principle but confused or misguided about what the work looks like in practice."

College students attending schools where legislation has been enacted are experiencing the impact with curricula changes and the elimination or restructuring or defunding of different identity support groups and services on campuses.

1.2.8 Belonging

Most of the empirical research connecting a sense of belonging to persistence is from the workplace. Ayre et al. (2013) found that in workplaces where there was an unusually high retention rate for female engineers, all the women revealed that they had come into the field strongly believing in themselves as engineers. The authors found that "the most striking features of the interviews were the confidence with which all the women asserted that they were good engineers, and that they belonged in engineering, although for some women it had taken time to feel they belonged (pp. 25–26)." Corbett and Hill (2015) argue that a sense of belonging is key to retention in any given field of study. There is a need for additional evidence from college-age populations.

Belonging can be enhanced by working on teams to solve problems. Programs like *Engineers Without Borders* (https://www.ewb-usa.org/) or student competition teams can help to build belonging, as long as the teams are sufficiently open and are not, themselves, a source of discrimination for women and students of color.

The focus of this work was on young women, but many of these factors are also correlated with young men's decisions to choose engineering or persist in engineering programs. In addition to these factors, another important factor that needs to be present for women is equal access to high-quality opportunities to develop a deep understanding of content and practices.

Precollege women need to have rich, robust, and scaffolded opportunities to engage with the core disciplinary ideas of engineering and the related science and engineering. As women develop proficiency with the engineering ways of thinking and the epistemic practices that generate engineering products, they become active participants in an engineering community and build their confidence in their abilities.

1.3 There Is a Place in Engineering for Everyone

There is an exciting opportunity for the engineering community to make the message of "Engineering is for Everyone" believable. Increasing interest in engineering among a diverse pool of students is the first step in diversifying the engineering workforce.

In 2008, the National Academy of Engineering published a report called Changing the Conversation (National Academy of Engineering, 2008). This report was the culmination of a study looking at the importance of messaging in recruitment of students to engineering. The study was a very targeted effort to find messages that would contribute to the public understanding, and appeal, of engineering. The study recommended that the engineering community engage in a concerted effort to reposition engineering in conjunction with the statement, "No profession unleashes the spirit of innovation like engineering. From research to real-world applications, engineers constantly discover how to improve our lives by creating bold new solutions that connect science to life in unexpected, forward-thinking ways. Few professions turn so many ideas into so many realities. Few have such a direct and positive effect on people's everyday lives. We are counting on engineers and their imaginations to help us meet the needs of the 21st century," and the tagline, "Engineers make a world of difference." These recommendations include shifting away from an emphasis on engineering as an application of math and science, which tends to cause students to self-select out.

Global Strategy Group and DiscoverE led a nationwide study in 2022 (DiscoverE, 2023) to examine whether the conclusions of Changing the Conversation held up with youth nearly 20 years later. The researchers found that high school students' interest in engineering grows with exposure to profiles of engineers and targeted key messages. When students see images and profiles of engineers who look like people they aspire to be, they are more receptive to the idea of pursuing a career in engineering.

The Messages Matter researchers indicate that the top message from the Changing the Conversation report, "Engineers can make a world of difference," was still very appealing. Parents and students in combination also found, "Engineering is a well-paid and prestigious field that sets students up for success," to be appealing, just as parents had done in the previous report. Finally, the idea, "Engineering is a career that is open to everyone," was judged to be appealing, if not necessarily believable.

Perhaps most impactful, however, is the finding that students still mostly associate building and math with engineering, and they see math skill as a gatekeeper for pursuing engineering in any form. Missing was any association of engineering with creative problem-solving or other mindsets common to the discipline.

Messaging is not the only issue impacting why there are so few women and people of color in the field. Opportunities to participate in meaningful, relevant precollege STEM vary greatly. All precollege youth need access to high-quality, hands-on experiences that introduce them to the role of engineering in our society, foster their substantive understandings of engineering knowledge, engage them with core disciplinary ideas, and develop their facility with engineering practices. As they do this kind of work, youth build their knowledge of and facility with engineering and view themselves as competent problem solvers and engineers, thus building engineering identities.

Families of these youth also need resources that help them to understand the possibilities that are available and pathways to a variety of STEM careers. Outreach events should invite families and communities to connect with engineering and science, recognize its role in their lives, and reflect upon their abilities to help shape future developments through civic action, coursework, and careers.

K-12 teachers, too, need deeper opportunities to become familiar and comfortable with engineering. Efforts should ensure that educators build their recognition about the diversity of perspectives, talents, and viewpoints that are needed to develop innovative solutions to current problems. Scaffolds to help educators articulate pathways to engineering work, highlight role models with similar interests and backgrounds, and nurture conversations and experiences that encourage females (and other students currently underrepresented in engineering) are needed.

Resources for guidance counselors are needed that help expand their visions of who should pursue an engineering career and become more knowledgeable about the pathways and programs that are available. Importantly, this expanded vision of who should be an engineer is also needed at the college and university level. From admissions counselors to engineering faculty and administrators, old mindsets that engineering is not appealing to women remain institutionalized in many ways.

1.4 Misconceptions

Misconceptions about women and engineering highlight the kinds of systemic change needed to change the underrepresentation of women in engineering. A few of those misconceptions are discussed below.

Girls Are Not or Should Not Be Interested in STEM
A simple Google image search on the term "toys for girls" in November 2023 produced 371 images of toys. Of those, one was related to science (growing crystals on a hedgehog). Sixty of the toys involved jewelry or makeup. Eighty involved animals (stuffed or animal care sets), and 170 involved dolls and playing house. Princesses were also a common theme. Of the 75 toys that did not fit one of the above categories, many were associated with phones and music.

A similar search for "toys for boys" yielded 155 toys related to vehicles, 21 weapons, 24 building sets, 15 robots, 12 action figures, and 23 sports-related items. Of the 81 in the "other" category, two were kitchen play sets, but none were related to dress-ups, dolls, jewelry/makeup, or animals. Ten were related to science (stomp rockets, slime making). Certainly, this search was not scientific. But it does illustrate a kind of societal assumption about what girls and boys like. As of 2024, Purdue University INSPIRE program has produced ten consecutive years of an Engineering Toy Gift Guide (https://engineering.purdue.edu/INSPIRE/EngineeringGiftGuide) that engage children and teens in engineering and design without link to gender.

Research shows that girls are indeed interested in science. In fact, in early grades (K-4 in the USA), girls are as interested in science as boys. Even in high school, girls perform equally well in mathematics courses, and they take them at similar rates. But in the USA, by the time girls get to high school, they take fewer physical science courses like physics. They take biology courses at a higher rate than boys (U.S. Department of Education, National Center for Education Statistics, 2000), which leads to misconception number two.

Girls Are More Interested in Life Sciences than Sciences
This misconception is tricky. Data do show girls gravitating to life science courses and even to engineering fields that are more closely related to life sciences, like biomedical engineering and chemical engineering. However, the reasons behind these tendencies are more complicated than simple preference. As stated in Baker and Leary (1995), "Both physical and biologic science are interesting to study in school, but physical science careers are avoided because they seem unrelated to the girls' concerns. Biologic science careers are perceived as helping people, animals, and the earth."

The actual preference is for relevance and an understanding of how the sciences help. We, as educators, scientists, and engineers, have not done a good job making the case for how science and engineering make a significant difference in the world. These results are consistent with the NAE study, "Changing the Conversation" (2008).

Lack of Ability
Surely, many of us remember the unfortunate quote by the President of Harvard University (Science, 2005) that women are underrepresented in tenured positions in science and engineering at top universities and research institutions because there are fewer women of high capability than men. There have been many responses to these remarks, but they articulated a societal belief rather succinctly: that women lack mathematical and scientific ability. Current data, however, do not support these conclusions. At one large, public university in 2023, women represent 33% of students in the College of Engineering and are retained and graduated at a rate approximately 10% higher than that of men. Lack of ability is not a contributing factor to underrepresentation of women in science and engineering.

In the early days of the space program, complex mathematical calculations were done by hand. The persons who did these calculations were largely women, as they were viewed as tedious and time consuming. The calculators, as they came to be called, were women of various ethnicities as well, as portrayed in the book, Hidden Figures (Shetterly, 2016).

Interestingly, however, mathematics ability has also been cited as a reason that fewer women pursue engineering degrees. A wide range of research has been conducted around this issue. Historical AP test data have shown men outscoring women on mathematics assessments, but this gender gap has shrunk in recent years. As of 2023, boys do still tend to outscore girls on achievement tests, like the AP, but girls actually outnumber boys in many math classes and have higher grades (Meiksins & Layne, 2023). Additionally, there is research that indicates that girls' scores on math tests are negatively impacted by asking about gender at the beginning of a test, i.e., stereotype threat (Spencer et al., 1999). This suggests that the tests, what they are measuring, or how they measure need further scrutiny. It is extremely likely that women do not lack mathematical ability, but rather have less self-confidence in mathematics, which can lead them not to pursue activities that they perceive as requiring significant mathematics.

Women Are Less Committed to Their Careers

Another misconception that many employers, including universities, hold about women in engineering and STEM is that women are not as committed to their careers. The contention is that women are more reluctant to spend long hours working and are more distracted by familial commitments. This misconception can be particularly insidious.

The number of women who graduate with degrees in engineering is significantly higher than those employed in engineering careers. This is likely due to several factors, including that women are frequently recruited for positions in sales or management due to their communication skills. Research done in 2016 showed that women did still take on a significantly larger role in family responsibilities than men, but even men were showing signs that they were becoming less willing to sacrifice a personal life for working long hours (Ecklund & Lincoln, 2016). Indeed, gender equity in engineering may benefit all genders by reducing the emphasis on long hours. But being unwilling to sacrifice a personal life in favor of work does not show lack of commitment. In one study of 3700 women engineers, reasons that women gave for their departure from engineering included spending more time with family, losing interest in engineering, a dislike for the engineering culture, and lack of opportunities to advance (Fouad & Singh, 2011). However, two-thirds of the women that left engineering were working in another industry, and almost a quarter of them held management or executive positions, thus suggesting a commitment to having a career in spaces where the work, opportunities, and environment supported their goals.

Girls and Women No Longer Experience Bias and Discrimination

A 2018 publication by the National Academies reported on sexual harassment of women in academic sciences, engineering, and medicine. The publication reported that incidents of sexual harassment defined to include "gender harassment (verbal and nonverbal behaviors that convey hostility to, objectification of, exclusion of, or second-class status about members of one gender), unwanted sexual attention (verbally or physically unwelcome sexual advances, which can include assault), and sexual coercion (when favorable professional or educational treatment is conditioned on sexual activity)" had not decreased over the previous 30 years (XE "Academia" National Academies of Sciences, Engineering, and Medicine, Policy and Global Affairs; Committee on Women in Science, Engineering, and Medicine, and Committee on the Impacts of Sexual Harassment in Academia XE "Academia", 2018). This is just one example of how bias and discrimination continue to impede the progress of women in STEM careers.

Interestingly, while gender discrepancies exist in the USA and Western Europe, this is not the case worldwide. In some countries in Eastern Europe, women outnumber men in science and engineering fields. Cultural practices (e.g., Soviet policies that tapped all talent) have led to societal conditions in which women in countries like Lithuania, Bulgaria, Latvia, Poland, and Serbia are equally

Fig. 1.2 Quote by Melinda Gates

"When we invest in women and girls, we are investing in the people who invest in everyone else." -
Melinda Gates

represented or outnumber men (Why Half the Scientists in Some Eastern European Countries Are Women, 2019). Such data highlight the societal and cultural origins of the discrepancies in the USA (Fig. 1.2).

Over the past few decades, dozens, if not hundreds, of papers, articles, and scholarly works have decried the imbalance of women in the field of engineering and have sought to explain the gap. *Despite all the challenges faced by women in engineering—the chilly climate, unequal pay, harassment, and worse—it is remarkable that so many women persist in the field of engineering or that so many young women even express an interest in becoming engineers.*

1.5 Conclusion: Looking to the Future

Women's place in engineering and in engineering education has come a long way, but it is clear that much more work needs to be done. The number of women in the engineering profession has remained static at 16% (this percentage is even lower for women of color) even with the initiatives designed and implemented for the past 20+ years to increase that number.

The USA is facing sharp declines every year in the number of high school graduates. Additionally, the population attending college is increasingly composed of women and other historically underrepresented groups that engineering struggles to attract. This is occurring at a crucial time as we face workforce needs due to retiring Baby Boomers, rise of new technologies, and global problems such as climate change, pandemics, and resource allocation, to name a few.

As this chapter demonstrates, clear evidence exists about how we can attract and retain women in engineering starting as early as elementary school. It is imperative that we continue to foster interest and positive attitudes about engineering, nurture self-efficacy, provide support networks, and create a sense of belonging. We need to continue to challenge misconceptions about women and engineering.

The fields of engineering are evolving and becoming much more multidisciplinary. Today's complex challenges need diverse perspectives to build solutions. The presence of women in engineering has had a tremendous influence on the climate of engineering education. The increasingly diverse students studying engineering present an opportunity to also increase the diversity of those educating these future engineers. Together, they will shape the future culture of engineering and engineering education.

The US industrial revolution brought us not only new technologies but also a system of problem-solving. Today, we recognize the need for more and more diverse minds to collaborate and solve current and future problems. Innovation and creativity, as well as empathy, lie at the heart of solutions to the difficult problems we face as a society. Engineering educators have a tremendous opportunity to influence the ways that we engineer, creating "empathetic engineers" that have the ability to devise sustainable solutions to global problems.

The importance of an inclusive mindset cannot be overstated. Problems are not solved by the same kinds of minds that created them. Increasing the participation of women and other groups traditionally underrepresented in engineering is one path to addressing the problems that challenge our societies and building a better world for everyone.

Acknowledgements The views presented in this chapter are those of the authors and not necessarily of the National Academy of Engineering and the National Academies of Sciences, Engineering, and Medicine.

References

Agbenyega, J. S. (2018). Examining early childhood education system in ghana: How can bourdieuian theorisation support a transformational approach to pedagogy? *International handbook of early childhood education* (pp. 673–690). Dordrecht: Springer Netherlands.

Andersen, L., & Ward T. J. (2014). Expectancy-value models for the STEM persistence plans of ninth-grade, high-ability students: A comparison between Black, Hispanic, and White students. *Science education 98*(2), 216–242.

Ayre, M., Mills, J., & Gill, J. (2013). 'Yes, I do belong': The women who stay in engineering. *Engineering Studies, 5*(3), 216–232. https://doi.org/10.1080/19378629.2013.855781

Baker, D., & Leary, R. (1995). Letting girls speak out about science. *Journal of Research in Science Teaching, 32*(1), 3–27. https://doi.org/10.1002/tea.3660320104

Barker, A. M. (2003). The effect of world war II on women in engineering. *Journal of Women and Minorities in Science and Engineering, 9*(3), 10. https://doi.org/10.1615/JWomenMinorScienEng.v9.i3.10

Bottomley, L. (2022). A Garden, Not a Pipeline. *ASEE Prism, 31*(7), 48.

Buontempo, J., Riegle-Crumb, C., Patrick, A., & Peng, M. (2017). Examining gender differences in engineering identity among high school engineering students. *Journal of Women and Minorities in Science and Engineering, 23*(3), 271–287. https://doi.org/10.1615/JWomenMinorScienEng.2017018579

Cech, E., Rubineau, B., Silbey, S., & Seron, C. (2011). Professional role confidence and gendered persistence in engineering. *American Sociological Review, 76*(5), 641–666. https://doi.org/10.1177/0003122411420815

Cole, D., & Espinoza, A. (2011). The postbaccalaureate goals of college women in STEM. *New Directions for Institutional Research, 2011*(152), 51–58. https://doi.org/10.1002/ir.408

Corbett, C., & Hill, C. (2015). *Solving the equation: the variables for women's success in engineering and computing*. American Association of University Women.

Dennehy, T. C., & Dasgupta, N. (2017). Female peer mentors early in college increase women's positive academic experiences and retention in engineering (S. Cheryan, Ed.). *Proceedings of the National Academy of Sciences, 114*(23), 5964–5969. https://doi.org/10.1073/pnas.1613117114
DiscoverE. (2019). *Despite the odds: Young women who persist in engineering*. DiscoverE.org/Despite-the-Odds.
DiscoverE. (2023). *Messages matter: Effective messages for reaching tomorrow's innovators*. DiscoverE.org/Messages-Matter.
Ecklund, E., & Lincoln, A. E. (2016). *Failing families, failing science*. Work-Family Conflict in Academic Science/NYU Press. Retrieved from https://nyupress.org/9781479843121/failing-families-failing-science/
Fouad, N. A., & Singh, R. (2011). Stemming the tide: Why women leave engineering. Research report University of Wisconsin, Milwaukee
Furter, W. F. (1980). *History of Chemical Engineering: Based on a Symposium Cosponsored by the ACS Divisions of History of Chemistry and Industrial and Engineering Chemistry at the ACS/CSJ Chemical Congress, Honolulu, Hawaii, April 2–6, 1979*. American Chemical Society.
Godwin, A., & Lee, W. C. (2017). A cross-sectional study of engineering identity during undergraduate education. Paper presented at 2017 ASEE Annual Conference & Exposition. Retrieved from https://peer.asee.org/27460
Gregory, S. L. (2015). African American female engineering students' persistence in stereotype-threatening environments: A critical race theory perspective. All Graduate Theses and Dissertations, Spring 1920 to Summer 2023. https://doi.org/10.26076/63aa-70ef
Griffith, A. L. (2010). Persistence of women and minorities in STEM field majors: Is it the school that matters? *Economics of Education Review, 29*(6), 911–922. https://doi.org/10.1016/j.econedurev.2010.06.010
Kim, J., Mayes, R. D., Hines, E. M., & Bryan, J. (2017). The parent–high school counselor relationship and students' postsecondary enrollment. *Journal of Counselor Leadership and Advocacy, 4*(2), 124–136. https://doi.org/10.1080/2326716X.2017.1322929
Lourens, A. S. (2014). *Engineering co-curricular role model interventions to develop women engineering students' self-efficacy at a South African university* (pp. 9–13). The International Academic Forum, European Conference on Education. Retrieved from https://papers.iafor.org/wp-content/uploads/papers/ece2014/ECE2014_01811.pdf
Maltese, A. (2008). Persistence in STEM: An investigation of the relationship between high school experiences in science and mathematics and college degree completion in STEM fields. PhD Dissertation. University of Virginia. Retrieved from https://www.researchgate.net/publication/234510692_Persistence_in_STEM_An_investigation_of_the_relationship_between_high_school_experiences_in_science_and_mathematics_and_college_degree_completion_in_STEM_fields
Maltese, A. V., & Tai, R. H. (2011). Pipeline persistence: Examining the association of educational experiences with earned degrees in STEM among U.S. students. *Science Education, 95*(5), 877–907. https://doi.org/10.1002/sce.20441
Marra, R. M., Rodgers, K. A., Shen, D., & Bogue, B. (2012). Leaving engineering: A multi-year single institution study. *Journal of Engineering Education, 101*(1), 6–27. https://doi.org/10.1002/j.2168-9830.2012.tb00039.x
Meiksins, P., & Layne, P. (2023). *Women in engineering: Analyzing 20 years of social science literature*. SWE Magazine. Retrieved from https://magazine.swe.org/lit-review-22/
Naphan, D. E. (2016). *Women, strategic identity management, and persistence in college engineering*. University of Nevada, Reno ProQuest Dissertations Publishing. Retrieved from https://www.proquest.com/docview/1808428021
National Academies of Sciences, Engineering, and Medicine, Policy and Global Affairs; Committee on Women in Science, Engineering, and Medicine, & Committee on the Impacts of Sexual Harassment in Academia. (2018). In F. F. Benya, S. E. Widnall, & P. A. Johnson (Eds.), *Sexual harassment of women: Climate, culture, and consequences in academic sciences, engineering, and medicine*. National Academies Press. https://doi.org/10.17226/24994

National Academy of Engineering. (2008). *Changing the conversation: Messages for improving public understanding of engineering*. The National Academies Press. https://doi.org/10.17226/12187

Ong, M. (2011). The status of women of color in computer science. *Communications of the ACM, 54*(7), 32–34. https://doi.org/10.1145/1965724.1965737

Pandey, E. (13 December, 2023). Anti-DEI movement expands in politics, business and academics. Axiox Politics & Policy. Retrieved from https://www.axios.com/2023/12/13/dei-programs-harvard-israel-gaza-protests

Paulsen, C. A. et al. (2021). Engaging Children and Families in Active, Environmental Science Learning through Digital Media. *International Journal of Early Childhood Environmental Education 8*(2), 43–58.

Raelin, J. A., Bailey, M. B., Hamann, J., Pendleton, L. K., Reisberg, R., & Whitman, D. L. (2014). The gendered effect of cooperative education, contextual support, and self-efficacy on undergraduate retention. *Journal of Engineering Education, 103*(4), 599–624. https://doi.org/10.1002/jee.20060

Rozek, C. S., Svoboda, R. C., Harackiewicz, J. M., Hulleman, C. S., & Hyde, J. S. (2017). Utility-value intervention with parents increases students' STEM preparation and career pursuit (S. T. Fiske, Ed.). *Proceedings of the National Academy of Sciences, 114*(5), 909–914. https://doi.org/10.1073/pnas.1607386114

Sadler, P. M., Sonnert, G., Hazari, Z., & Tai, R. (2012). Stability and volatility of STEM career interest in high school: A gender study. *Science Education, 96*(3), 411–427. https://doi.org/10.1002/sce.21007

Sahin, A., Ekmekci, A., & Waxman, H. C. (2018). Collective effects of individual, behavioral, and contextual factors on high school students' future STEM career plans. *International Journal of Science and Mathematics Education, 16*(Suppl 1), 69–89. https://doi.org/10.1007/s10763-017-9847-x

Science. (2 February, 2005). Harvard President Summers' Remarks About Women in Science, Engineering

Shetterly, M. L. (2016). *Hidden figures: The American dream and the untold story of the black women mathematicians who helped win the space race*. HarperCollins.

Spencer, S. J., Steele, C. M., & Quinn, D. M. (1999). Stereotype threat and women's math performance. *Journal of Experimental Social Psychology, 35*(1), 4–28. https://doi.org/10.1006/jesp.1998.1373

Tancredi-Brice Agbenyega, E. (2018). *"We are fighters": Exploring how Latinas use various forms of capital as they strive for success in STEM*. Temple University ProQuest Dissertations Publishing. https://doi.org/10.34944/dspace/3616

Trenor, J. M., Yu, S. L., Waight, C. L., Zerda, K. S., & Sha, T.-L. (2008). The relations of ethnicity to female engineering students' educational experiences and college and career plans in an ethnically diverse learning environment. *Journal of Engineering Education, 97*(4), 449–465. https://doi.org/10.1002/j.2168-9830.2008.tb00992.x

U.S. Department of Education, National Center for Education Statistics. (2000). In Y. Bae, S. Choy, C. Geddes, J. Sable, & T. Snyder (Eds.), *Educational equity for girls and women NCES 2000–030*. U.S. Government Printing Office. Retrieved from https://nces.ed.gov/pubs2000/2000030.pdf

Why Half the Scientists in Some Eastern European Countries Are Women. (18 July, 2019). *The Economist*. Retrieved from https://www.economist.com/europe/2019/07/18/why-half-the-scientists-in-some-eastern-european-countries-are-women#

Wigfield, A. (1994). Expectancy-value theory of achievement motivation: A developmental perspective. *Educational psychology review 6*(1), 49–78.

Young, J., & Young, J. (2018). The structural relationship between out-of-school time enrichment and black student participation in advanced science. *Journal for the Education of the Gifted, 41*(1), 43–59. https://doi.org/10.1177/0162353217745381

Chapter 2
Women in Engineering Programs and WEPAN

Beth M. Holloway, Jane Z. Daniels, Susan Staffin Metz, and Carol B. Muller

2.1 Introduction

To go fast, go alone. To go far, go together.—Unattributed African Proverb

The power of community to catalyze change is shown nowhere better than through the development of Women in Engineering (WiE) programs and WEPAN. A symbiotic relationship between those programs at colleges and universities and the WEPAN organization characterized WEPAN's beginning in 1990 through its growth for the next three decades, exemplifying the idea that building community can leverage individual efforts to create a multiplicative effect. WiE programs and initiatives have been and continue to be important and impactful community-building efforts to support and encourage gender equity in engineering and as a force for positive change on individual campuses. Bringing together both established and nascent programs and initiatives in higher education, along with advocates and allies from industry, government, and foundations, creates broader impacts and a force for change around the nation and beyond. WEPAN was established as a result of the collective interest in creating a network to support, sustain, and ensure continuous learning for those who seek to advance gender equity in engineering.

As authors of this chapter, we emphasize that we are writing from our collective experiences and references, an approach which has limitations. Over the years,

B. M. Holloway (✉) · J. Z. Daniels
Purdue University, West Lafayette, IN, USA
e-mail: holloway@purdue.edu

S. S. Metz
Stevens Institute of Technology, Hoboken, NJ, USA
e-mail: susan.metz@stevens.edu

C. B. Muller
Stanford University, Stanford, CA, USA
e-mail: cbmuller@stanford.edu

J. P. Carpenter (ed.), *Women in Academia*, Women in Engineering and Science,
https://doi.org/10.1007/978-3-031-97938-5_2

through the inclusive, collaborative organization that WEPAN was designed and continues to be, there are many, many histories, levels and roles of engagement, perspectives, and experiences of various individuals that may differ from ours, and this description we provide cannot possibly capture the richness of the contributions and engagement of hundreds of people and organizations over time. In taking up the slightly daunting request to write about the development of WiE programs and WEPAN, we recognize that there are many varied and diverse experiences and leadership perspectives that have contributed to the evolution of WiE programs, and to WEPAN's strength, resilience, versatility, mission, evolution, and success over the years, far more than could be described in one book chapter. Given the inclusive, distributed, operational nature of WEPAN as an organization, from which its collaborative and generous culture grows, this chapter describes only a few of the leaders and initiatives that have emerged from this remarkable organization.

2.2 The Development of Women in Engineering Programs

In 1970, women comprised less than 1% of graduates of engineering programs and participants in the engineering workforce; gradually, particularly as labor market needs grew, employers became more aware of the potential economic advantage of tapping this source of intellectual talent to develop future engineers. The supply of engineers graduating from engineering schools had not yet experienced any substantial increase from women students. At the same time, attitudes about women working professionally, and the roles open to them, were undergoing significant change beyond the conventional fields, such as teaching, nursing, and secretarial work. As a result of the civil rights and women's movements, some of the previous barriers to women's employment in various professions fell with the Equal Pay Act of 1963, and in 1972, Affirmative Action (Executive Order 11246) and Title IX of the Educational Amendments Act aimed to eliminate discrimination based on sex, and persistent feminist activists promoted use of these laws and provided data to encourage policy and programs to improve women's opportunities and abilities to break into formerly all-male or heavily male-dominated areas of study and work. Consequently, more women considered careers outside the home and across a broader spectrum of opportunities.

In 1969, with visionary action, Purdue University's President Frederick L. Hovde and Head of Freshman Engineering Harold Amrine hired Donna Frohreich McKenzie to be the coordinator for a program for women in engineering. This concern for the dearth of women in engineering and a desire to bolster engineering enrollments at Purdue built on earlier efforts. As a land grant institution, Purdue graduated its first woman with an engineering bachelor's degree in 1897—Martha Dicks Stevens, in civil engineering—but the enrollment of women in the engineering program was sparse. In the 1930s and 1940s, both Dr. Lillian Moller Gilbreth and Amelia Earhart were employed by Purdue University to counsel Purdue's women students (Daniels & LeBold, 1982). In 1935, Earhart became a visiting faculty member at Purdue as an advisor to aeronautical engineering and a career counselor to women students (Goldstein & Dillon, 1997; Kass-Simon et al., 1993). When Gilbreth was hired at Purdue, she became the nation's first woman professor of Mechanical Engineering (Chong & Proctor, 2021). By 1940, as a full professor, Gilbreth divided her time among Purdue's departments of industrial engineering, industrial psychology, home economics, and the dean's office, where she counseled women on careers (Gugin & Clair, 2015; Kass-Simon et al., 1993; Graham, 1998). In 1947, the Gamma Chapter of Pi Omicron, an honorary society for women in engineering, was established on the Purdue campus. In 1954, 5 years after the first meeting of the national Society of Women Engineers (SWE) and 2 years after SWE was incorporated, Purdue's Pi Omicron organization officially changed its name to the Society of Women Engineers, Purdue Student Section. By 1969, the formation of a WiE program was the next logical step to encourage women into and through its engineering program. The initial focus of the program was in promoting visibility of engineering as a career to high school women and encouraging enrollment in Purdue Engineering. While the number of women studying engineering increased substantially, there were significant differences in retention and graduation of women engineering students relative to men; therefore, additional activities were developed to support current women students and promote community.

Stevens Institute of Technology in Hoboken, New Jersey, was founded in 1870 as a private, all-male, mechanical engineering degree-granting university. Women did attend its graduate school; Beatrice Hicks, co-founder and first president of SWE, graduated from Stevens with an M.S. degree in 1949. It was not until 1971, however, a year that many US universities incorporated coeducational study into their programs, that Stevens admitted its first 19 women undergraduates. Established in 1976, Stevens' chapter of SWE officially chartered 30 student members—almost the entire women student body. Growth in the number of women undergraduates was very slow at a time when enrollment was declining. To increase the number of women undergraduates and generate a pipeline to engineering, Dr. Edward Friedman, Dean of the College, supported an effort to launch a series of pre-college programs. The goal of the programs was to introduce high school women to engineering and science career opportunities due to a widespread lack of understanding of these types of careers among students, parents, counselors, and teachers. The growth of these programs continued, funded by corporations and the National Science Foundation. Stevens created an Office of Women's Programs in 1980 to conduct research and implement pre-college- and college-level programs to increase public understanding of engineering and recruit and retain women in engineering and science and hired Susan Staffin Metz to lead this office. These initiatives, funded in large part by corporations, corporate foundations, and the National Science Foundation, expanded significantly, leading to the establishment of the Lore-El Center for Women in Engineering and Science in 2002, renamed the Lore-Center for Leadership, a living-learning center on the Stevens campus.

In the late 1980s, the WiE Initiative was established at the University of Washington under the leadership of the Dean of Engineering, Dr. J. Ray Bowen. Dr. Suzanne Brainard was hired in 1987 as the inaugural director of WiE. Suzanne's strengths in evaluation were used from the beginning to design and implement WiE's programs that were based on research and evaluation of their effectiveness. Programs were initially developed to focus on the retention of undergraduate and graduate women students in the College of Engineering. Recognizing that women in many of the science disciplines were encountering the same retention challenges, Washington's WiE programs expanded to the Center for Women in Science and Engineering (WISE) in the mid-1990s. Other characteristics somewhat unique to the University of Washington's WISE program were the director's significant involvement in international organizations, the development of programs that were sensitive to the issues important to international women students, and the enrichment of existing programs by including international women students as members of the WISE staff.

In recognition of their work with women in engineering students, all three of these WiE-focused programs (at Purdue University, Stevens Institute of Technology, and the University of Washington) have received the Presidential Award for Excellence in Science, Mathematics and Engineering Mentoring (PAESMEM) through the White House Office of Science and Technology Policy.

2.3 The Uneven Growth of WiE Programs

Throughout the United States, the number of recruitment and retention activities for women in engineering at US colleges and universities increased from the early 1970s through the 1980s. Other early programs began at the University of Michigan, Ohio State University, and Rutgers University. In their early years, however, little or no cross-institutional communication existed between these programs; their activities resulted in reinvention of programs and initiatives, sometimes inefficient use of available funding, insufficient program evaluation, and, ultimately, a slowing of progress toward educational equity for women in engineering, as these programs faltered in strength. A comparison of information from the 1982 and 1987 WiE Directory of College/University Programs indicates a decline in all areas of programming for women engineering students except for the number of schools offering monetary awards to first-year undergraduate and graduate women students. By 1989, women were earning 53% of all bachelor's degrees conferred in the United States but only 39% of the bachelor's degrees conferred in math and science (excluding social sciences and psychology) and only 15% of the engineering bachelor's degrees awarded (National Research Council, 1992). By 1990, there were about 10 formally organized WiE Programs in the United States (Anderson-Rowland et al., 1999), though there was continued interest in increasing the numbers of women in engineering. For example, a number of national and even international conferences were held in the 1980s focused on women and/or minoritized students in engineering and science, such as the "Recent Studies and New Directions in Research Concerning Women in Science and Engineering" conference, convened by the National Research Council in 1981.

2.4 The Creation of WEPAN, the Women in Engineering Programs & Advocates Network

The increased interest in starting or expanding WiE programs and activities in the United States created both excitement and concerns for the directors of programs at schools that had early success. They were balancing the management of their own programs while working with colleagues at other institutions who needed their advice. With Jane Daniels' (Purdue University) and Susan Metz's (Stevens Institute of Technology) experience in directing successful WiE programs, and Suzanne Brainard's (University of Washington) grant-writing expertise and knowledge of funding sources, the trio approached the director of the Career Access Program at the National Science Foundation (NSF) in 1989 to ascertain interest in funding a national conference of individuals interested in the recruitment, admission, retention, and graduation of women engineering students. They hoped to encourage engineering schools throughout the United States to establish effective programs for women in engineering or to expand the scope of existing programs in the most effective way possible.

A formal request for NSF funding was based on the results of a questionnaire sent to engineering deans, women engineering faculty members, and Society of Women Engineers student section advisors and counselors determining their interest in a national conference focused on WiE programs. The majority of engineering schools accredited by the Accreditation Board for Engineering and Technology (now referred to as ABET) responded positively and committed to sending a representative to such a conference. With a grant secured from NSF and supplemented by funds from ten corporations, the first WEPAN conference was held on May 30–June 1, 1990, in Washington, DC.

Highlight: The Abstract for the First NSF Grant Launching WEPAN

This project involves a cooperative effort between Purdue University, the major institution, and the University of Washington and Stevens Institute of Technology. It proposes to provide greater access for women to careers in engineering while counteracting projected shortages of engineering personnel by assisting and encouraging engineering schools throughout the United States to establish innovative programs for women in engineering or to expand the scope of existing programs. Specifically, this program would (1) organize a national conference to study issues and present models of programs for women in engineering; (2) establish the first national network of Women in Engineering Program Administrators; (3) provide technical and programmatic assistance to institutions desiring to initiate, replicate, or expand Women in Engineering programs in a cost-effective manner; (4) establish a central source of research, information, and resource materials about women in engineering and related areas; and (5) disseminate information nationally through conferences and publications. The heart of the proposal begins with items (1) and (2)-the holding of the national conference and the establishment of the national network. In this regard, the three institutions, Purdue, Washington, and Stevens, will have key administrators participating and coordinating—Purdue, the entire program; Washington, the western region; and Stevens, the eastern region. A timetable has been developed, and the conference will be held in October, two months after the planning sessions both years. Subsequently, annual conferences and continued upgrading of the directories might be possible. Because of the work of Dr. Daniels at Purdue, the network is fairly well in hand, and definitive were given for its strengthening, collecting, and dissemination of information. The project will provide a very meaningful source of information, and to the extent that the network is successful, it should provide a very useful conduit for women to know about, recognize, and enter careers in engineering. Financial support will come from eight sources: the three sponsoring institutions, various engineering organizations, conference participants, industry, the AAAS, and, hopefully, the National Science Foundation.

Two hundred individuals attended the conference representing a wide variety of organizations and interests: WiE program administrators; engineering department heads; deans of engineering and their representatives, and a wide variety of other academic leaders, including those at community colleges; faculty members; high school math and science teachers; and representatives of corporations, government labs and agencies, private foundations, corporate foundations, engineering professional societies, and other nonprofit organizations. Workshops and presentations were developed to be widely appealing and inclusive. Sessions pertaining to starting a program, pre-college outreach, retention programs, reentry programs, program evaluation, funding opportunities, graduate-level programs, faculty issues, and issues of intersectional diversity were offered. The final sessions of the conference included regional meetings of attendees interested in continuing a network of advocates of women in engineering programs. The most conclusive mandate emerging from these meetings was ***"A new organization is needed which specifically addresses methods to attract and retain women in engineering."***

Based on this mandate, in 1990, Suzanne Brainard, Jane Daniels, and Susan Metz established WEPAN[1] as a 501(c) (3) nonprofit organization. They also created three regional centers for the administration of the organization, with Purdue University as the home of the Midwest Regional Center, Stevens Institute of Technology as the home of the Eastern Regional Center, and University of Washington as the home of the Western Regional Center. Table 2.1 lists the first WEPAN Board of Directors. Instrumental to the early development of WEPAN was the continuing support of the founders' home institutions and staff members. Cathy Deno at Purdue University was particularly valuable to the success of WEPAN's organizational stability and annual conference administration. She served as Director of Member Services from 1990 to 2013. Along with Tricia Bailey at Stevens Institute and Claire LeBlanc at the University of Washington, Cathy coordinated registration for each annual conference and coordinated printing of the

---

[1] WEPAN was originally established as the "Women in Engineering Program Administrators Network"; the name was subsequently changed to "Women in Engineering Programs & Advocates Network" in 1991 and later changed to "Women in Engineering ProActive Network" in 2008.

Table 2.1 The first WEPAN Board of Directors, 1990–1991

Name	Affiliation
Jane Daniels—President	Purdue University[a]
Susan Metz—Vice President	Stevens Institute of Technology[a]
Suzanne Brainard—Secretary	University of Washington[a]
Betty Vetter—Treasurer	Commission on Professionals in Science and Technology
Howard Adams	GEM Consortium
Mary Anderson-Rowland	Arizona State University
Eleanor Baum	Cooper Union[a]
Marilyn Berman	University of Maryland
George Brewster	Corning Inc.[a]
Deborah Grubbe	E. I. DuPont
Gloria Hill	Carnegie Mellon[a]
Margie Kintz	Intel
Marsha Matyas	American Physiological Society[a]
Harold Shields	ALCOA
Myrna Whigham	Iowa State University

[a]Indicates participation in the 1990 Conference Planning Committee

Conference Proceedings. She also administered WEPAN-L, an electronic communication network.

Highlight: The Impact of the First Conference

In 1990, Dr. Karan Watson was an associate professor at Texas A&M University (TAMU) and was also serving in her first administrative position: assistant dean in charge of the Honors Program, Minority Engineering Program, and Graduate Program. When her dean received information about the first WEPAN conference, he brought it to her attention and asked her to attend. Other than her own experience as an engineering undergraduate, graduate student, and newly tenured faculty member, Karan had no formal knowledge of gender issues or programming directed at women students in engineering. Two experiences of that 1990 conference stand out in her memory more than 30 years later. First was the willingness of everyone present to share their experiences, both in formal sessions and in social interactions throughout the conference. And second was a breakout session for women faculty members. The session for women faculty began, like many gatherings of her peers, with complaints about the difficulties the women faculty members were experiencing with both their colleagues and their students. Yet, at this session, one of the more experienced faculty members listened in a supportive way but then redirected the conversation. The rest of the session focused on the value women faculty members brought to the conference and what they could do on their individual campuses to support one another and women engineering students.

Karan also remarked that she appreciated that WEPAN could articulate differences between WiE programs and Women in Science programs. "We could learn from them (Women in Science Programs) but develop programs that were appropriate for engineering students." Karan said that she left that first conference with new information, good data, and a great network of like-minded colleagues. When Karan returned to TAMU, she requested and was provided the resources to create a WiE program, whose first director, Jan Rinehart, soon also became involved in WEPAN's leadership. Karan recalls several reasons why she engaged in the leadership structure of this new organization. "My father always told me that if I saw value in the mission of an organization, I should help shape the future of that organization." In addition, "WEPAN did a phenomenal thing—giving directors and others a voice on their campuses on behalf of women in engineering. It helped program administrators understand the importance of getting access to disaggregated university data; reminding them that programs needed to be based on data and developed with evaluation results to be credible to engineering faculty and administrators. WEPAN was invested in growing their impact; the organization maintained its roots (supporting WiE program directors) while developing in new ways to change the system." Karan served on WEPAN's Board of Directors from 1991 to 2001[2] and as its Secretary from 1993 to 1996 and Treasurer from 1996 to 2001; she eventually increased her levels of responsibility at TAMU, including Provost and Executive Vice President.

2.5 The Development of WEPAN

As WEPAN took shape, the organization developed sources of funding and revenue streams, including those from private and corporate foundations, corporate sponsorships, membership dues, conference fees, and government agencies. Though these foundations, corporations, and agencies are too numerous to list, the decision not to list them in no way is meant to discount the enormous importance of such support for WEPAN and its initiatives. In addition, many of the grants for WEPAN's projects were supported through the universities of the founders and others, so that essential functions such as sponsored project administration, accounting, compliance, payroll, benefits, and other human resource activities could be supported by these universities through their indirect cost recovery. The vast majority of WEPAN's activities and initiatives have been funded through grants, without which WEPAN could not have delivered on its critical and strategic mission to provide professional development and a community of practice for WiE program administrators and women faculty, to support and generate research on women in engineering, to support pre-college awareness of engineering, and to advocate for systemic change in engineering education.

To recognize and promote the good work of its members, and to provide examples of excellence, WEPAN started an awards program in 1996 with three strategic awards: an award for an exemplary WiE program, an award recognizing significant achievement in research in areas related to women in engineering, and an award for extraordinary service to the organization. The awards program has expanded over the years and now includes 17 awards, based on WEPAN's continuing strategic emphases.

[2] Karan Watson related this story to Jane Daniels via an interview conducted in December 2023.

2.5.1 WEPAN Professional Development and Community Building Initiatives

One of the foundational professional development and community building initiatives was the WEPAN annual conference. Held every year between 1990 and 2017, the annual conference was an opportunity to welcome new researchers and practitioners and to share research and best practices. Each conference had a theme that was driven by the organization's focus areas and informed by the status of engineering culture. As an example, the theme of the 1999 conference was "Moving Beyond Individual Programs to Systemic Change" (Metz, 1999). Starting in 1997, WEPAN partnered with the National Association of Multicultural Engineering Program Advocates (NAMEPA) to offer joint conferences every several years. This partnership was born out of WEPAN's commitment to diversity and inclusivity. Conference proceedings were published for each conference to inform the broader engineering community about the work of WEPAN members, later archived at Penn State University through the leadership of Barbara Bogue, a former WEPAN board member and active leader in the organization.

Starting in 1992, *WEPAN Regional Training Seminars* were designed to provide resources and expertise to women in engineering program directors, who each often served as the sole person responsible for improving the representation of women in engineering at their institutions. In total, about 200 teams from colleges and universities participated in 2.5-day workshops led by a WEPAN leadership team and offered throughout the United States. In addition to funding participation in the seminars, WEPAN secured external resources to develop and test curriculum to support the workshops, which resulted in the development and implementation of nearly 100 new campus-based efforts. The resulting publication, *Increasing Access for Women in Engineering*, with Susan Metz as editor, included chapters and supplementary materials to develop pre-college programs, recruitment and retention programs, climate assessment, and successful fundraising efforts. A second publication, *A Curriculum to Train Mentors and Mentees* (Brainard, 1998), included an administrator's guide, handbooks for participants, a bibliography of resources, an evaluation module, a video of scenarios depicting mentoring relationships, and a guidebook for facilitating group discussions. These curricula, funded by the Fund for the Improvement of Postsecondary Education (FIPSE), were disseminated to every engineering school in the United States.

In 1997, WEPAN hosted the *Achieving Success in Academia* conference to begin to address the low representation of women faculty in engineering in the United States. Funded by the National Science Foundation, this conference offered strategies to enhance participants' careers in academia and provide insight into successfully navigating the tenure track. Participants included 34 non-tenured, tenure-track women engineering faculty and 28 women engineering graduate students interested in academic careers from 28 different institutions. Each participating graduate student was accompanied by a faculty member from their institution to help initiate or maintain a mentoring relationship. Participating faculty shared challenging experiences at their universities as they pursued tenure and were in search of answers. Dr. Denice Denton, then dean of engineering at the University of Washington, gave a riveting presentation appreciated by all. Her candor, direction, challenging questions, and sharp responses to concerns raised by the participants were the highlights of the event.

WEPAN, in partnership with Penn State University, developed new initiatives to increase the number of women and other underrepresented groups in faculty positions in engineering and science. Through the *Faculty for the Future* website, universities had the ability to post jobs for faculty and research positions, postdocs, graduate school fellowships, and undergraduate summer research opportunities. In addition, undergraduate and graduate students could submit and update their resumes and send their resumes to universities seeking candidates.

Highlight: WEPAN as a Supportive Community

Planning the Puerto Rico conference and then attending were one of the highlights of my time on the WEPAN board. I first got involved with WEPAN when as an Assistant Professor, I had really bad few days (i.e., journal article rejected, NSF proposal not funded, and just feeling overwhelmed). I remember that it was really late at night, and I emailed Susan Staffin Metz just to share my frustration and she saw that I needed a "lifeline." She sent me the most kind, concerned, and inspiring response I could have ever imagined. After that, I was totally sold on a WEPAN membership and later joined the board. WEPAN is so much about supporting the "whole woman" in an engineering career. That was the reason I got involved, joined the board, and stayed involved for so many years—Pamela R. McCauley, PhD, Dean, School of Engineering, Widener University.

Clearly, at least in part, as a result of WEPAN programs and initiatives, the number of WiE programs increased to approximately 85 in 2002, at which time WEPAN was recognized with the Presidential Award for Excellence in Science, Mathematics and Engineering Mentoring (PAESMEM) through the White House Office of Science and Technology Policy.

With these and other initiatives, a strong cadre of diverse experts and successful WiE program directors was cultivated, and those individuals continued to volunteer and lead WEPAN in subsequent decades. Highlighting one example of how WEPAN leaders extended the impact and national voice of WEPAN beyond the organization itself and leveraged their experience in their career trajectories, past WEPAN presidents have assumed various roles after their terms ended, such as leaders of other STEM organizations, leadership roles in higher education, various types of advisory boards, and principal or co-principal investigators on grant funding.

Highlight: WEPAN Incubates a New Venture—MentorNet

As assistant dean at Thayer School of Engineering, in early 1990, I had co-founded a campus-wide project focused on the retention of undergraduate women in engineering and sciences at Dartmouth College. I attended the 1990 Women in Engineering conference, meeting and learning from so many there, and returned to my home campus with influential data, key strategies, and important contacts to further our goals. E-mail was becoming ubiquitous on campus at that time; it facilitated communication with students about the project, and its events and activities, and provided a means for informal mentoring and encouragement for large numbers of students. Among many different activities, through an industrial site visit involving just a few hours' interaction with working engineers, students came with new

insights about possible career opportunities, the kinds of study that might be useful in preparing for such careers, and new positive motivation and energy. They met role models. They met people who enjoyed and were stimulated by their work, who were having fun, engaged in team-based projects and activities, and obviously excited about their endeavors. The power of interactions with scientific and technical professionals in industry was plain, yet too few students had such opportunities, and they were complicated to arrange. I began to think about linking students with engineers and scientists in industry more readily via e-mail relationships and, in 1993, outlined a concept paper about forming a large-scale electronic mentoring program pairing students with mentors in industry, enriched by engaging a multitude of students, universities, industrial professionals, corporations, and professional societies. That prospectus led to support for a 2-year pilot program at Dartmouth, to test and further develop the concept with the idea that a large-scale initiative could be developed through WEPAN as a national program. The pilot "e-mentoring program" began in 1995, and evaluation helped to improve and measure the program's success.
The initial evaluation led to further grant support the next year to develop a strategic plan for implementing a national program, by then dubbed "MentorNet." The plan in turn led to start-up grants through WEPAN to build a team and the needed resources, including a new partnership with the College of Engineering at San José State University and the San José State University Foundation; to launch the national program in early 1998, initially pairing 225 students from 15 universities with mentors from 93 different employers; and to build and develop a technology-based platform to support formerly labor-intensive programs at a large scale. Over the next 10 years, 30,000 pairs of mentors and protégés from more than 200 institutions of higher education were matched in one-on-one mentoring relationships. By 2000, with encouragement from its advisory board and in consultation with the WEPAN president and board of directors, MentorNet became a separate nonprofit 501(c) (3) organization in the State of California and continued to collaborate with WEPAN, regularly offering partner meetings, workshops, and other sessions at WEPAN's annual conference and introducing its participants and partners to WEPAN. In 2001, MentorNet was recognized with the Presidential Award for Excellence in Science, Mathematics and Engineering Mentoring (PAESMEM) through the White House Office of Science and Technology Policy. Though I opted to move on to other work in higher education in 2008 after 12 years of developing and leading MentorNet, the organization and its basic one-on-one program continue. In 2014, MentorNet's leadership arranged for MentorNet's one-on-one program to be sustained as a program of Great Minds in STEM—Carol B. Muller, PhD, Founder and former CEO, MentorNet.

2.5.2 WEPAN Initiatives Focused on Pre-college Programming

As part of the organization's efforts to spread awareness of careers in engineering, WEPAN developed several initiatives meant to inspire pre-college students, support those administrators who worked with pre-college populations, and promote further efforts in this space. Examples of two such initiatives follow.

In 2000, WEPAN was awarded a grant from Lucent Technologies to develop materials to help change engineers' public image among those who matter most, the engineers of the future. The goal of Making the Connection was to create an awareness and interest about engineering among students in grades 3–12, with a particular emphasis on girls, young women, and other underrepresented minority students. A series of 15 hands-on engineering-based activities, a presenter's guide, and age-appropriate newsletters were developed by experts for five educational levels: middle elementary, upper elementary, middle school, early high school, and late high school. The curriculum was designed to highlight the contributions engineers make to improve the quality of lives; to dispel stereotypes of engineering as a field for "white males"; to make the connection between students' math and science courses and engineering; and to explore the skills and interests needed to become an engineer.

In the summer of 2000, WEPAN managed a pilot project at six US universities. *Exploring Interest in Technology and Engineering (EXITE)* was a weeklong summer program designed to expose middle-school girls to engineering and computer science. Women engineers from IBM, university professors, and graduate students served as project teachers, speakers, and mentors. Each university's program format was similar but tailored to highlight specific areas of expertise. Among the subjects covered in seminars were teamwork, problem-solving, decision-making, and communicating effectively. Laboratory experiences included animated web-page design, computer-chip design, laser optics, computer hardware, and more. WEPAN member universities participating in the pilot project were Purdue University; University of Puerto Rico, Mayaguez; University of Texas at Austin; University of Michigan, Ann Arbor; Binghamton University (New York); and Georgia Institute of Technology.

Highlight: WEPAN's Role in the Creation of Introduce a Girl to Engineering Day

I managed IBM's participation with WEPAN, where I met Susan Metz and Jane Daniels, and Carol Muller, then leading MentorNet. This network of women became my network of support as I evolved IBM's Women in Technology program. IBM was scheduled to sponsor the 2001 National Engineers Week (NEW), an annual event hosted by the National Society of Professional Engineers, with the purpose of increasing understanding of and interest in engineering and technology careers. There is much research that confirms that improved familiarity with engineering has a direct impact on both consideration of engineering as a career as well as on the number of college students pursuing engineering degrees. IBM was a long-time supporter of NEW, and as part of the planning for EWeek 2001, I was put in touch with Leslie Collins, the executive director of NEW, to see what we might do in terms of women engineers as part of the 2001 program. At the annual WEPAN conference, I introduced Leslie to Susan and Carol, and we brainstormed thoughts on what we might do. With enthusiasm, we landed on the idea for Introduce a Girl to Engineering Day. When we brought the idea back to both IBM and NEW, we quickly gained their support, and Introduce a Girl to Engineering Day was launched! At IBM, the excitement within the female technical community spread from location to location where teams of IBM women engineers and computer scientists participated. At the IBM location in Austin, Texas, for example, there were 92 participants in the first Girl Day. That particular event has become Girl Day at the University of Texas at Austin, which has hosted over 8500 K-8th graders each year—Linda Scherr, Program Director, IBM Women in Technology (retired).

2.5.3 Research as a WEPAN Priority

From its very beginnings, WEPAN prioritized research on women in engineering and their experiences. In addition to the published proceedings of the annual conferences, WEPAN and its members conducted research and published literally thousands of publications about women in engineering.

One such study was the *WEPAN Pilot Climate Study*. In 1998, 29 WEPAN member institutions participated in a survey developed by WEPAN (Metz et al., 1999). The goal was to assess engineering student perceptions of the academic climate at colleges and universities in the United States. The survey focused on seven major areas: quality of teaching, quality of teaching assistants, quality of labs, quality of departmental assistance, general questions about engineering, levels of self-confidence in academic courses, and demographic data. Pilot sites were able to add several questions that could be tailored to their institution. Analyses of the data, from over 8000 responses, focused particularly on gender differences. Each participating university received a report of their findings, shared it with academic leaders, and used the data to support institutional change initiatives. One important finding from the study was that women and men perceived the undergraduate engineering experience differently and women's perceptions were less positive.

By 2007, WEPAN recognized the need for a knowledge center to enable researchers and practitioners to locate resources important to their work quickly and easily, without reinventing programs, initiative, and research or relying on outdated information. Such information was sometimes difficult to locate as it had become very extensive over the years and was published across many diverse organizations, websites, educational institutions, and individual project sites. The *WEPAN Knowledge Center* (WKC) was launched in 2009, as part of a project funded by the National Science Foundation. The WKC was subsequently rebranded as the Women in STEM Knowledge Center, knowing that root causes of women's underrepresentation span STEM fields and recognizing the multidisciplinary impact of gender in STEM issues. The Women in STEM Knowledge Center is broadly available to share knowledge center information and resources with the larger STEM community.

Another example of WEPAN's significant focus on research is the WEPAN Betty Vetter Award for Research, which recognizes notable achievement in research related to women in engineering. The award is named in memory of Betty M. Vetter, longtime director of the Commission on Professionals in Science and Technology and WEPAN's first treasurer. The honorees for this award, listed in Table 2.2, are an esteemed group of researchers.

Table 2.2 Honorees of the WEPAN Betty Vetter Award for Research

Year	Name	Affiliation
1996	Emily M. Wadsworth	Purdue University
1998	Pat Campbell	Campbell-Kibler Associates, Inc.
2000	Elaine Seymour	University of Colorado at Boulder
2002	Mary Frank Fox	Georgia Institute of Technology
2001	Eleanor Babco	Commission on Professionals in Science and Technology
	Elizabeth Cannon	Calgary University
	Valarie Haines	Calgary University
	Jean Wallace	Calgary University
2003	Sue Rosser	Ivan Allen College, Georgia Institute of Technology
2004	Judy L. Lupart	University of Alberta
	M. Elizabeth Cannon	University of Calgary
2005	Sheryl Sorby	Michigan Tech
2006	Virginia Valian	Hunter College
2007	Barbara Bogue	The Pennsylvania State University
	Rose Marra	University of Missouri, Columbia
2009	Karen Tonso	Wayne State University
2010	Peggy Meszaros	Virginia Tech
	Elizabeth Creamer	Virginia Tech
2011	Michelle Hebl	Rice University
2012	Daryl E. Chubin	American Association for the Advancement of Science (AAAS)
2013	Matthew Ohland	Purdue University
	Susan Lord	University of San Diego
	Michelle Camacho	University of San Diego
	Catherine Brawner	Research Triangle Educational Consultants
	Richard Layton	Rose-Hulman Institute of Technology
	Russell Long	Purdue University
	Mara Wasburn	Purdue University
2014	Amy Sue Bix	Iowa State University
2015	Mary Anderson-Rowland	Arizona State University
2016	Mary Besterfield-Sacre	University of Pittsburgh
2017	Julie P. Martin	Clemson University
2018	Lynnette Madsen	National Science Foundation
2019	Joyce B. Main	Purdue University
2022	Ethel Mickey	University of Massachusetts
	Ramón Barthelemy	University of Utah
2024	Brooke Coley	Arizona State University

Highlight: Hiring the First Executive Director

From WEPAN's inception, the AT&T Foundation was a steadfast supporter and partner and propelled WEPAN's growth, culminating in a collaborative arrangement where WEPAN assumed fiscal responsibility for managing AT&T graduate student fellowship funds. This financial entrustment was instrumental in sustaining WEPAN's operations, marking a significant milestone in 2004 with the hiring of Diane Matt as its first Executive Director. This development allowed the organization to move beyond relying exclusively on volunteers and piecemeal grant-funding through universities. Diane's contributions to WEPAN were significant, as she positioned the organization for professional sustenance and set up structures for organizational governance and finance. An experienced nonprofit executive, Diane worked thoughtfully with the Board of Directors to help it develop to meet its fiduciary and governance responsibilities, meet challenges, create policy, and onboard new members. She became the primary point of contact for WEPAN, communicating WEPAN's interests and mission to its many constituents, in writing and in speaking (see for example Matt, 2007), and representing WEPAN and its interests to a wide variety of audiences. During her tenure as WEPAN's executive director, in collaboration with others, Diane developed and wrote grant proposals for complex and far-reaching new initiatives and worked to implement those initiatives, further extending the visibility and reach of WEPAN. Over the years she worked with WEPAN, she gained the trust and appreciation of numerous WEPAN board members, members, funders, and constituents, with whom she built strong and lasting relationships, benefiting WEPAN and its mission.

2.5.4 A Focus on University Culture and Systemic Change

In addition to an organizational focus on research and publications from the very beginning of the WEPAN organization, there has also been a focus on understanding university culture (e.g., the *WEPAN Pilot Climate Study*) and advocating for systemic change. The foundational work of Seymour and Hewitt's (1997) longitudinal cohort research long ago indicated that women and men earn similar grades in engineering courses, and women who leave engineering have higher grades than men who stay. "It is not, therefore, poor academic performance that drives women out of engineering, but higher levels of dissatisfaction" (Adelman, 1998). WEPAN's advocacy to resist framing women as "deficient," and its focus on changing the culture of engineering so that all can thrive, started as one of several areas of focus for the organization, but eventually became its central focus, under which all other priorities connect. WEPAN established itself as a prominent and respected voice within the national landscape. An example was the organization's invitation to participate in the 2002 Pan-Organizational Summit on the US Science and Engineering

Workforce, convened by the National Academies. This honor placed WEPAN among 32 leading professional organizations shaping the discourse on this critical issue. WEPAN's submitted position paper articulated the need for improving public awareness of engineering as a profession and systemic change in the university culture which taught engineering and included specific recommendations. Many of the final recommendations from the Summit included those identified by WEPAN (Fox, 2003). WEPAN also participated in a Capitol Hill press conference in 2005 encouraging lawmakers to leverage Title IX with respect to women in STEM.

2.5.5 The Integration of the Four Frames for Gender Equity into WEPAN

In 2005–2007, through a strategic planning process, the WEPAN Board of Directors decided to focus the attention of the organization on the culture of academic engineering. This shift centered the mission of the organization on the need for systemic culture change in engineering, instead of being one of several aspects of its mission. The newly developed mission statement became one of being a catalyst, advocate, and leading resource for institutional and national change that enables the success of all women in engineering. The shift was an important one for WEPAN, as it used organizational change as an overarching priority under which all other priorities resided. But organizational culture and systemic change were also conceptually difficult to talk about and define, and there were many questions about how WEPAN's work of supporting WiE program directors and the work of WiE programs themselves fit into the mission. In 2011, Beth Holloway brought the Four Frames for Gender Equity (Ely & Meyerson, 2000) to WEPAN as a way to help the organization talk about culture change and the role of supporting women where they are. The "Four Frames" became a lens through which WEPAN and its members could envision, create, view, and evaluate efforts to promote gender equity that went beyond "fixing the women." This framework brought common language related to gender equity efforts to the WEPAN community, especially important as more engineers and others not steeped in social science research became leaders of WiE programs and initiatives (see highlight below). The connection of the Four Frames to WEPAN brought together for the first time the research on gender in organizations with the dearth of women in engineering and offered a visual map and frame of reference. While each of the Four Frames has limitations (sometimes problematic), WEPAN chose to move forward with them as they clearly outline the ways in which gender equity will not be achieved without culture change.

Frame 1: Equip the Women This frame is foundational to most campus-based WiE programs, whose activities are designed to support women in the current culture of engineering. WEPAN supports WiE programs through sharing research and best practices related to efforts and activities of this type.

Frame 2: Create Equal Opportunities Through this frame, WEPAN focuses on advocacy for policy and practice changes within organizations and institutions to ensure that women have equal access to opportunities for development and success, including sharing research and approaches that can be adapted to campuses and corporate environments.

Frame 3: Value Differences This frame affirms and values the differences between women and men, leading to an understanding that what women bring to the profession is both valuable and needed.

Frame 4: Revise Culture This frame sets up culture change as the goal so that all people can thrive and propel their organizations to new levels of success. WEPAN supports its members in becoming advocates for and leaders of culture change within their institutions through leadership and professional development.

Several WEPAN initiatives have focused on understanding and changing culture in engineering academia. One, for example, was ENGAGE (NSF Award # 0833076), an NSF Extension Service Project modeled on the successful Cooperative Extension Service program at state land grant institutions developed under the leadership of Dr. Jolene Jesse, program officer at the National Science Foundation. Led by principal investigator Susan Metz, and co-principal investigators Diane Matt and Dr. Patricia Campbell (Campbell-Kibler Associates), ENGAGE, an 8-year multimillion-dollar initiative, was launched in 2009. ENGAGE provided resources to implement scalable, relatively low investment (in time and money) strategies substantiated by a rigorous body of research that supported the benefit of their use for increasing engineering student engagement and retention. Strategies focused on the first 2 years of college, when engineering students, particularly those who are underrepresented in engineering, are most at risk to change majors. The strategies included improving and increasing interaction between faculty and students, using everyday examples that are familiar to students to teach technical concepts, and improving students' spatial visualization skills. ENGAGE worked closely with 30 partner engineering schools to understand opportunities and barriers to implementation of the three strategies. EngageEngineering.org proved critical in not only supporting implementation among ENGAGE partner schools but also reaching the broader engineering community. Between 2011 and 2017, resources for implementing each strategy were downloaded almost 350,000 times and accessed by over 140,000 visitors, demonstrating wide reach across engineering schools beyond ENGAGE's partner schools and implying broader implementation of ENGAGE strategies.

Another example was *TECAID—Transforming Engineering Culture to Advance Inclusion and Diversity*. This engineering education research project focused on implementing an intensive, 2-year department-level leadership development initiative for five selected mechanical engineering department leadership teams and had an overall goal to build the capacity of the department leadership to create, develop, and sustain an inclusive department culture. Three specific professional development areas were targeted: change leadership, engineering culture, and inclusive pedagogies. The broader significance and importance of this project were in its ability to positively impact department culture resulting in an educational environment that welcomes all students, particularly those from underrepresented groups. The project utilized evidence-based interventions to effect a positive organizational change, increasing the awareness of factors that negatively impact students and changing the classroom environment to facilitate improved student learning. TECAID was proposed and implemented in partnership with the American Society for Mechanical Engineering (ASME), further demonstrating WEPAN's dedication to collective community building.

Highlight: A Perspective as an Engineer Moving into WiE Leadership …

My experiences as a first-generation engineering student in the late 1980s and early 1990s and as a practicing engineer throughout the 1990s created a desire to become part of the system that educated me, in the hopes of improving those experiences for others. Guided by my experiences in corporate recruiting and as a supervisor and mentor for newly graduated engineers and bolstered by my project management skills and systems thinking, I felt that I could make a difference in the education of women engineers. While I had first-hand experience with the engineering curriculum and the culture of engineering, my knowledge base was lacking in a number of important ways. I had limited knowledge of the research about women in engineering, organizational culture, and culture change and limited knowledge of proven best practices for inspiring and supporting women into and through their engineering education. I joined WEPAN for access to the research and best practices about effective ways to support current engineering students, engage future engineering students, and advocate for change. Through WEPAN, I gained that knowledge, and I came to understand that my experiences were but one representation of the experience of being a woman in engineering, and they were not wholly representative of all experiences of women in engineering. WEPAN also provided me with a community of practice and gave me opportunities to develop my academic leadership skills. I grew my professional network through WEPAN and have leveraged that network throughout my academic career—Beth M. Holloway, PhD, Senior Assistant Dean for Student Access and Success and the Leah H. Jamieson Director of WiE, College of Engineering, and Professor of Engineering Practice, School of Mechanical Engineering, Purdue University.

2.6 WiE Programs Today

As the Women's Experiences in College Engineering (WECE) report states, "The size and focus of WiE programs vary. In general, programs offer academic and social support for female engineering undergraduates. However, program directors may also recruit women to apply to the engineering school, conduct outreach to students in grades K-12, and provide support for graduate students and faculty women in engineering. WiE program directors have created a wide range of activities and resources to encourage and support women who are considering or pursuing an engineering major" (Goodman, 2002). Additionally, some WiE programs also conduct research on gender in engineering, or on their own programs, in addition to evaluation of their activities.

WiE programs vary structurally with respect to where they administratively reside inside an institution: within a college, at an institution-wide level, or within a particular department. The staffing levels of WiE programs also vary significantly, from programs staffed by a single (or even partial) person to those that have many staff members. Some WiE initiatives are administered in conjunction with initiatives focused on other underrepresented identities in engineering, under the same department umbrella and sometimes sharing staff. Programs and activities vary between institutions, shaped by local context and needs. While many programs focus solely on engineering, some programs combine engineering with science or STEM more broadly. WEPAN continues to be the organization that brings together WiE programs and initiatives. Figure 2.1 indicates some of the reasons that WiE programs engage with WEPAN.

Fig. 2.1 Common reasons for WEPAN engagement

2.7 WEPAN Today: Strength Through Partnerships

In 2018, toward the end of its third decade, WEPAN partnered with the American Society for Engineering Education (ASEE)'s Minorities in Engineering Division (MIND) and Women in Engineering Division (WIED) and the National Association of Multicultural Program Advocates (NAMEPA) to create the *Collaborative Network for Engineering and Computing Diversity* (CoNECD). As a conference, its purpose was to provide a forum for exploring current research and practices to enhance diversity and inclusion of all underrepresented groups in engineering and computing professions. With an initial 3-year commitment to engage its members through CoNECD rather than host its own conference, WEPAN's leadership experimented and engaged yet again with a unique collaboration. At the same time, WEPAN increasingly took advantage of improving online technologies to support virtual meetings and webinars to engage its broad array of constituent members, including those whose financial support might not allow them to attend conferences in person.

Highlight: The Foundations of CoNECD

Dr. Bevlee Watford joined WEPAN after being hired as the founding Director of the Center for the Enhancement of Engineering Diversity (CEED) at Virginia Polytechnic Institute and State University. She needed access to best practices and research about how to best support underrepresented group in engineering, and WEPAN was able to provide information about what to do and why those initiatives were effective. In addition, WEPAN provided Bevlee with a network, connecting her with many others who were working on similar issues within their own institutions. Bevlee was also deeply engaged in NAMEPA and ASEE and had served as president of all three organizations at various points in time. She was convinced that all three groups would be stronger if they worked together and was instrumental in creating the WEPAN/NAMEPA joint conferences. By 2018, Bevlee conceived of and led the creation of the Collaborative Network for Engineering and Computing Diversity (CoNECD) as a conference dedicated to all the diverse groups that comprise the engineering and computing workforce. Asked about the influence that WEPAN had on her, she said, "WEPAN is my discipline—if it was one. WEPAN encompasses recruiting, retention, and broadening participation, and is my community. After all these years, I am still learning from the organization and fellow members."[3]

2.7.1 WEPAN Connects with the ARC Network

In late 2020, WEPAN's Board of Directors chose Dr. Ershela Sims to lead WEPAN as its executive director. Shortly thereafter, the ARC (ADVANCE Resource and Coordination) Network, an ambitious initiative founded under the leadership of Dr.

[3] Bevlee Watford related this story to Beth Holloway via an interview conducted in December 2023.

Heather Metcalf in 2017, joined forces with WEPAN, recognizing both WEPAN's position as the leading champion in North America to propel the inclusion of women in the field of engineering and the highly aligned synergies of their respective goals, values, and cultures. Dr. Metcalf then became WEPAN's Director of Research, and with WEPAN as the backbone organization of the ARC Network, the affiliation has strengthened the abilities of both the ARC Network and WEPAN to develop and thrive going into the future. The ARC Network builds upon the history of the National Science Foundation's ADVANCE program, working toward gender equity for faculty in higher education science, technology, engineering, and mathematics (STEM) disciplines.

Characterized as "the STEM equity brain trust," the ARC Network "recognizes achievements made, while producing new perspectives, methods, and interventions with an intersectional, intentional, and inclusive lens," and works with the WEPAN community to provide engagement through the ADVANCE Equity in STEM Community Convening, bringing together strategic change-makers across a variety of sectors who are focused on systemic change in STEM, and with programs such as a Virtual Visiting Scholars program and emerging research workshops.

As the ARC Network moved to join WEPAN, it further deepened and strengthened WEPAN's long history of practice and interest in recognizing the intersectional perspectives needed to advance gender equity in engineering and related fields. Dr. Stephani Page, originally the ARC Network's community engagement manager, became WEPAN's Director of Strategic Initiatives. In this role, she continues to develop WEPAN programs, now including virtual Women of Color Summits and Program Days, and WiE Program Days, developing and serving those supporting women in engineering on university campuses. Dr. Page also takes the lead in creating partnerships and connections for other major initiatives furthering WEPAN's goals, such as the Women of Color in Engineering Collaborative (WCEC). Through these efforts, WEPAN builds on its roots to consider how power structures and avoidance of discomfort still reinforce systemic inequities.

2.8 Conclusions

As of this writing in 2024, WiE programs and broader DEI efforts on campuses are facing some significant challenges that make a national community of advocates and allies even more critical. Many WiE programs and related programs and initiatives recently have been subjected to Title IX challenges, several state legislatures have begun to restrict DEI efforts on college campuses, and some institutions themselves through their governing bodies have also prohibited or restricted DEI efforts. Though some WiE programs continue to thrive, and new efforts continue to be started, others have been reorganized or disbanded due to a myriad of campus reasons. The connection of WiE program administrators to each other and to research, through WEPAN, has never been more critical.

WEPAN is well poised for the future to continue to leverage networks of advocates who are supporting and leading organizational change. WEPAN's vision is to continue to be a resource and community of practice for WiE programs, to continue to collect and disseminate research and best practices for inclusive practices and cultures through the Women in STEM Knowledge Center, and to be a resource, advocate, and source of expertise for creating broadly inclusive cultures in both academia and industry. There have been so many contributors to WEPAN's development along the years, and its distributed and inclusive approach to leadership and engagement has meant that literally hundreds of people have made major contributions to its structural, cultural, programmatic, strategic, and financial development over the years and decades since its founding. Our chapter in this book can only skim the surface of WEPAN's history and briefly highlight but a tiny fraction of its extensive initiatives. WEPAN has always been, and continues to be, a highly collaborative, inclusive organization whose members build on their learning from and relationships with one another to create incredible value in the service of advancing equity in engineering.

Highlight: The Future of WEPAN

When thinking about the future of WEPAN, there is still work for us to do. I am optimistic because of the impact I see that we are making. I have observed the respect that WEPAN has as an organization because of its collaborative focus, inclusive values, talented staff and leadership, and committed board and membership. One of the things that I am most proud of is the work that we are doing to support and collaborate with our fellow STEM professional societies—Ershela Sims, PhD, Executive Director, WEPAN.

References

Adelman, C. (1998). Women and Men of the Engineering Path. In *A Model for Analyses of Undergrad Careers, Office of Educational Research and Improvement*. US Department of Education.

Anderson-Rowland, M., Blaisdell, S. L., Fletcher, S. L., Fussell, P. A., McCartney, M. A., Reyes, M. A., & White, M. A. (1999). A collaborative effort to recruit and retain underrepresented engineering students. *Journal of Women and Minorities in Science and Engineering, 5*.

Brainard, S. G. (1998). *A Curriculum for Training Mentors and Mentees*. Women in Engineering ProActive Network.

Chong, I., & Proctor, R. W. (2021). Lillian Gilbreth and Amelia Earhart: How an Eye Toward Diversity Brought Two Pioneers Together. *Ergonomics Design, 29*, 13–18.

Daniels, J., & LeBold, W. (1982). Women in engineering: A dynamic approach. *Effective Strategies for Increasing Participation of Women in Science*, pp. 139–163

Ely, R. J., & Meyerson, D. E. (2000). Theories of Gender in Organizations: A New Approach to organizational Analysis and Change. *Research in Organizational Behavior, 22*, 103–151.

Fox, M. A. (2003). *Pan-Organizational Summit on the US Science and Engineering Workforce: Meeting Summary*. Government-University-Industry Research Roundtable (US), 2003 North Carolina State University (p. 201). National Academies Press.

Goldstein, D. M., & Dillon, K. V. (1997). *Amelia: The Centennial Biography of an Aviation Pioneer*. Potomac Books Inc.

Goodman, I. F. (2002). Final Report of the Women's Experiences in College Engineering (WECE) Project. Online Submission

Graham, L. (1998). *Managing on her own: Dr. Lillian Gilbreth and women's work in the interwar era*.

Gugin, L. C., & Clair, J. E. S. (2015). *Indiana's 200: The People Who Shaped the Hoosier State*. Indiana Historical Society.

Kass-Simon, G., Farnes, P., & Nash, D. (1993). *Lillian Moller Gilbreth and the Rise of Modern Industrial Engineering*. Indiana University Press.

Matt, C. D. (2007). WEPAN History and Current Update. In *2007 ASEE Annual Conference & Exposition, Honolulu, Hawaii*.

Metz, S.S. (1999). Welcome Address. Retrieved from https://journals.psu.edu/wepan/article/view/58094/57782. Accessed 15 Aug 2024

Metz, S. S., Brainard, S., & Gillmore, G. (1999). National WEPAN Pilot Climate Survey Exploring the Environment for Undergraduate Engineering Students. In *1999 International Symposium on Technology and Society-Women and Technology: Historical, Societal, and Professional Perspectives. Proceedings*. Networking the World (Cat. No. 99CH37005) (pp. 61–72). IEEE.

National Research Council. (1992). *Science and engineering programs: On target for women?* The National Academies Press.

Seymour, E., & Hewitt, N. M. (1997). *Talking About Leaving*. Westview Press.

Chapter 3
The Society of Women Engineers Fills a Need

Jill S. Tietjen

3.1 Introduction

As the disciplines of engineering evolved, so did the associated need for engineering education. As institutions of higher learning did not admit women in the 1700s and 1800s (and sometimes the 1900s), women did not have access to an engineering education. Once women could be educated as engineers, the next hurdle was becoming a member of the professional engineering societies. As with engineering education, these organizations were not welcoming to women either. Although to a slight extent during the World War I era and to a much greater extent during the World War II era women were encouraged to become educated and employed as engineers, the educational arena and the workplace were not so interested in women once the war ended. However, after World War II, the women themselves were no longer content following societal norms and expectations. They founded the Society of Women Engineers in 1949–1950 to encourage women to be engineers and provide professional development opportunities for those in the profession.

3.2 Engineering Education

As the engineering disciplines evolved in Europe in the 1700s and 1800s, formal schools were established to teach the needed curriculum. European universities, however, did not admit women. Most European universities had been founded to teach theology, medicine, and law; these professions were all closed to women. So, women, with very few exceptions, were denied access to a university-level education. The École Nationale des Ponts et Chaussées (National School of Bridges and

J. S. Tietjen (✉)
Technically Speaking, Inc., Greenwood, CO, USA

J. P. Carpenter (ed.), *Women in Academia*, Women in Engineering and Science,
https://doi.org/10.1007/978-3-031-97938-5_3

Roads) was established in 1747 in France as the first formal school of engineering (Ambrose et al., 1997). In the nineteenth century, the practice and focus of engineering was significantly expanded with the development of Newtonian mechanics and the development of the steam engine. These developments led to an even greater need for formal education (Kass-Simon & Farnes, 1990).

Male French engineers assisted the colonies during the American Revolution, stayed to construct the new nation, and provided the foundation of engineering faculty for the U.S. Military Academy (USMA) at West Point (Barker, 1994). When it was established in 1802 to educate engineers, the USMA did not admit women. The Rensselaer Polytechnic Institute, established in 1824 in Troy, New York, now the oldest surviving nonmilitary engineering school in the USA, also did not admit women (Ambrose et al., 1997; Turner, 1954). Prior to the Civil War, there were only six schools of engineering across the entire USA; none admitted women (Turner, 1954). The women who called themselves engineers during this time were not formally educated in an engineering degree program.

Ellen Henrietta Swallow Richards (Fig. 3.1) was one of these nontraditional engineers. Although she was not an engineer by training, she contributed much to the establishment of the forerunners of environmental and sanitary engineering and

Fig. 3.1 Ellen Henrietta Swallow Richards (Courtesy Wikipedia)

is credited as the woman who founded ecology and home economics. When she applied to the chemistry department at the Massachusetts Institute of Technology (MIT) for a graduate degree in chemistry in 1870, she was not accepted because the department did not want its first graduate degree to go to a woman. She was not rejected, however (as she was at all the other universities where she had applied), but instead was allowed to enroll as a candidate for a second bachelor's degree. She was classified as a special student who did not have to pay tuition (she had already received a bachelor's degree from Vassar College). Richards did not know that MIT had admitted her without tuition so that they could deny she was officially enrolled if anyone complained. She was awarded a bachelor's degree in chemistry from MIT. MIT formally admitted women in 1878 and awarded its first advanced degree in 1886 (Kass-Simon & Farnes, 1990; Rossiter, 1992).

While the US population was centered in the East, the colleges in the West and mid-West formally admitted women earlier than East Coast institutions, primarily because many state-supported institutions (land grant institutions) were established as a result of the 1862 Morrill Act (Turner, 1954). But even then, the number of women formally studying engineering in the late 1800s was very few. The first woman to graduate from an engineering degree program in the USA was Elizabeth Bragg, who completed a B.S. in civil engineering at the University of California at Berkeley in 1876 (Kass-Simon & Farnes, 1990). Kate Gleason (Fig. 3.2) was the first woman to enter Cornell's Sibley College of Engineering in 1884, but did not stay to complete her studies, as she was called back to help the family business (Barker, 1994). Perhaps the second female engineer by education, Elmina Wilson graduated in 1892 from Iowa State College with a civil engineering degree and was the first female instructor at that school (LeBold & LeBold, 1998). In 1893, Bertha Lamme graduated from The Ohio State University with a degree in mechanical engineering with an option in electricity. She was the first woman to graduate with a degree in a field other than civil or architectural engineering (Kass-Simon & Farnes, 1990).

When the Society for the Promotion of Engineering Education (later named the American Society for Engineering Education) was formed in 1893, only the three women noted above were recorded as having received engineering degrees in the USA. However, as women began to enter the educational system, graduate, and then try to find work as engineers, a backlash developed (LeBold & LeBold, 1998).

Fig. 3.2 Kate Gleason (Courtesy Wikipedia)

3.3 Professionalization

Professionalization of engineering began in the late nineteenth and early twentieth centuries. As women were finally able, at least in small numbers, to gain an engineering education and engineering employment, they also endeavored to join the engineering societies. These organizations, however, did not welcome women and developed a strict hierarchy of requirements for each of several levels of membership. They were already in the midst of upgrading themselves, and the entrance of women into the profession was not seen as a positive development by most of the men involved in the leadership positions of these organizations (Rossiter, 1992).

Professionalization, in this case, meant upgrading the membership or image of a profession by excluding or diminishing the influence of persons who could be perceived to be "amateurs." Professionalization in engineering included deliberately creating barriers between engineers with college degrees and relevant professional experience and those other "engineers" who had learned their jobs by experience and lacked the "professional" credentials. In professional societies, professionalization often meant raising the standards of membership and led to great concerns about the perceived prestige of the organization. As most engineering schools did not admit women (and thus women could not get the desired "professional" credentials), the most significant impact of professionalization was to exclude women (Rossiter, 1992).

Professionalism was probably also a by-product of the state of engineering education. Engineering educational standards in the late 1800s and early 1900s were not yet at a level necessary to earn a college education. As a consequence, engineers were not invited to serve on national scientific advisory boards, nor were engineers recognized as part of the established scientific community until 1916. And it was not until 1932 that the Engineering Council for Professional Development (ECPD), now known as ABET, was created to provide accreditation of engineering degree programs, in partial response to reports sponsored by the engineering societies (Barker, 1994).

3.4 Professional Societies

By the end of the nineteenth century, civil and mechanical engineering were firmly established as engineering disciplines, with electrical and chemical engineering following closely behind (Rossiter, 1992). Engineering societies were forming. These types of organizations are deemed the hallmark of a profession. They define intellectual style and norms of conduct and generally act to promote the interests of their members. The early engineering societies placed a high value on free enterprise, individualism, hard work, ambition, and success intertwined with the concepts of prestige, status, and professionalism (Barker, 1994).

The "Founder" societies, the five original engineering societies that founded the United Engineering Society in 1904 (which later became the United Engineering Foundation), included the American Society of Civil Engineers (ASCE), the American Institute of Mining Engineers (AIME) (now called the Society for Mining, Metallurgy and Exploration—SME), the American Society of Mechanical Engineers (ASME), the American Institute of Electrical Engineers (AIEE—a predecessor organization to the Institute of Electrical and Electronics Engineers—IEEE), and the American Institute of Chemical Engineers (AIChE) (United Engineering Foundation, 2000a, b; IDIS, 2000a). A brief look at their history and their admittance (or more accurately, their lack of admittance) of women shows the impact that professionalization and the associated membership requirements had on the recognition for and advancement of women in the engineering profession.

ASCE, America's oldest national engineering society, was founded in 1852. Twelve founders met at the Croton Aqueduct in New York City on November 5, 1852, and agreed to incorporate as the American Society of Civil Engineers and Architects (later the ASCE) (ASCE, 2000a, b). Emily Warren Roebling, probably the first woman field engineer, became the first woman to address the ASCE in 1892, when she argued that her husband should not be replaced as the formal director for the construction of the Brooklyn Bridge (Kass-Simon & Farnes, 1990). In 1909, Nora Blatch de Forest, a graduate of Cornell University in the top five of her class, was admitted as a "junior member" of the ASCE but was unable to advance any higher. When her junior membership expired in 1916, ASCE refused to promote her to full membership, despite her meeting the stated requirements, and, instead, dropped her from the rolls. She brought a lawsuit against the Society but did not prevail (Rossiter, 1992). Elsie Eaves became an associate member of ASCE in 1927 and later the first female member (1957), first female life member, first female Fellow, and first woman elected to honorary membership (1979) (Kass-Simon & Farnes, 1990; SWE, 1959).

AIME was founded in 1871 by 22 mining engineers in Wilkes-Barre, Pennsylvania (IDIS, 2000a, b). The first woman member was Ellen Henrietta Swallow Richards, who became the first woman member of any engineering society when she was elected a full member of AIME in 1879. Richards was aided by her MIT degree, her publications in mineral chemistry, and the fact that her husband was a vice president of the organization (Rossiter, 1992). In 1917, the Woman's Auxiliary to the AIME was established and is still active today as the WAAIME (IDIS, 2000b; Society for Mining, Metallurgy and Exploration, 2015).

By 1880, 85 engineering colleges had been established in the USA, and most offered a full mechanical engineering curriculum. Thirty engineers met in New York City in February 1880 and decided to form the ASME. In April, a formal organizational meeting was held with 80 engineers at the Stevens Institute of Technology in Hoboken, New Jersey. The first annual meeting of the organization was held in November 1880 (ASME, 2000). The first woman member, Kate Gleason, was admitted to full membership of the ASME in 1914 (Ingels, 1952). Lydia Weld, a 1903 naval architecture graduate of MIT, became an associate member of ASME in 1915. She was allowed to become a full member in 1935, when the ASME finally granted full membership status to women (LeBold & LeBold, 1998).

By 1884, twenty-five prominent figures in electrical technology signed a "call" to establish an American electrical engineering national society, mindful that civil, mining, and mechanical engineers had already established their own national societies. Twenty-five electrical engineering practitioners met in the headquarters of the ASCE on April 15 to devise an organizational structure for what became the AIEE. The first general meeting was held on May 13, also at ASCE headquarters (IEEE, 2000a). In 1926, Edith Clarke, who would later become one of the first AIEE female fellows, was the first woman to address the AIEE (Read & Witlieb, 1992). As late as 1942, there were only three women in the AIEE and more than 17,000 men (Read & Witlieb, 1992). The Fellow grade was established in 1912 for engineers who had demonstrated outstanding proficiency and had achieved distinction in their profession. It was not until 1948, however, that the first women were elected AIEE Fellows (Rossiter, 1995). These three distinguished women were Edith Clarke, who significantly contributed to knowledge about and modeling for electric utility systems; Vivien Kellems, the founder of Kellems Company, a manufacturer of cable grips and shell lifters; and Mabel Rockwell, who significantly contributed to electrical control systems (IEEE, 2000b; Goff, 1946; SWE, 1999).

AIChE was founded in 1908 at the Engineers' Club in Philadelphia by 19 men. Chemical engineering was just coming into its own and was somewhere between chemistry and mechanical engineering. The founding of AIChE helped to establish chemical engineering as a separate discipline. At the time, the founding members of AIChE believed that about 500 people were practicing chemical engineering across the country (AIChE, 1999). The first female member of AIChE was Margaret Hutchinson Rousseau. The first woman to receive an Sc.D. in chemical engineering from MIT and who made significant contributions to the field was not admitted until 1945 (Rossiter, 1995).

In addition to the Founder societies, an engineering honor society, Tau Beta Pi, was established. The engineering equivalent of Phi Beta Kappa (which had been founded at the College of William and Mary in 1776), Tau Beta Pi, was established at Lehigh University in 1885. It was founded to:

> *mark in a fitting manner those who have conferred honor upon their alma mater by distinguished scholarship and exemplary character as undergraduates in engineering, or by their attainments as alumni in the field of engineering, and to foster a spirit of liberal culture in engineering colleges.*

Membership in Tau Beta Pi was limited to men until 1969. "Women's Badges" had been authorized in 1936 as an alternative to membership for women. Only 619 Women's Badges had been awarded by 98 chapters until women were admitted to full membership in 1969, 84 years after the founding of the organization (Tau Beta Pi, 1999; Phi Beta Kappa, 2000).

Other engineering societies, in addition to the Founder societies, came into existence in the early to mid-1900s. However, these organizations also excluded women from membership or often relegated them to lower membership status. Marie Luhring was elected as an associate member of the American Society of Automotive Engineers in 1920 (Ingels, 1952). That same year, Ethel H. Bailey became the first full female member of the organization (Ingels, 1952). The American Society of Safety Engineers, founded in 1911 in New York City with 62 members as the United Society of Casualty Inspectors, admitted its first female member, Vera Burford, in 1946 (Rossiter, 1995; ASSE, 2000). The first time a woman was elected as a junior member of the Society of Naval Architects and Marine Engineers was 1946 (U.S. Government Printing Office, 1948). The first female fellow of the Illuminating Engineering Society, Gertrude Rand, a researcher on the way color perception is affected by illumination and on color blindness, was elected in 1954 (Read & Witlieb, 1992; Rossiter, 1995).

Lillian Gilbreth (Fig. 3.3), "the first lady of engineering" and the co-founder of the field of industrial engineering, was made an honorary member of the Society of Industrial Engineers in 1921 (as a personal favor to her husband, Frank Gilbreth), but not admitted to regular membership (Rossiter, 1992: IIENet, 1999a). She was, however, the first woman elected to the National Academy of Engineering (NAE), an event that occurred in 1965, only 1 year after the founding of the NAE in 1964 (NAE, 2000; IIENet, 1999b).

Fig. 3.3 Lillian Moller Gilbreth (Photograph courtesy of Society of Women Engineers Records, Walter P. Reuther Library, Wayne State University)

Once women were admitted to full membership, it would still be decades before these professional societies elected their first woman president. That occurred in 1986–1987, when Nancy Fitzroy (Fig. 3.4) was elected president of ASME, becoming the first woman to lead any major engineering society. Next was IEEE. Martha Sloan became its first female president in 1993. In 2003, Patricia Galloway became the first woman president of ASCE as did Diane Dorland for AIChE. Barb Filas served as the first female president of SME in 2005 (Wastewater Digest, 2003; ASME, 2014; AIChE, 2024; IEEE Computer Society, 2024; SME, 2024; University of Arizona, 2023).

Fig. 3.4 Nancy Fitzroy (Photograph courtesy of Society of Women Engineers Records, Walter P. Reuther Library, Wayne State University)

3.5 World War I

During World War I (April 1917–November 1918), women were encouraged to participate in the workforce and support the war effort. This "war to end all wars" has also been called the first engineers' war, as the results of military applications of scientific and technological advances appeared as tanks, airplanes, and submarines (Barker, 1994). Because of the shortage of male engineers during World War I, women had the opportunity to work in factories and offices where they were instrumental in keeping the manufacturing industry working (LeBold & LeBold, 1998). They were drawn into tool design and chemical research and designed buildings and automobiles (Goff, 1946).

However, because the USA was involved in the conflict for only 19 months, the manpower emphasis was on the emergency training of mechanics to service the new military technology and not the education of engineers with degrees (Barker, 1994). Some women became wartime mechanics through special training; some of those women, because of this introduction, graduated with engineering degrees in the 1920s. Approximately 120 women, some without any college degree, worked during the war in engineering jobs. The end of the war signaled the end of this opportunity. However, prior to 1920, at least 45 women are known to have graduated from college with engineering degrees (Turner, 1954). It is ironic that war, an activity that women tend to oppose, increases women's economic progress, as men are drawn out of the civilian workforce (Goff, 1946).

Many historians believe that the vote for women's suffrage, which was finally ratified in 1920, was out of gratitude for women's efforts during the war (Rossiter, 1992). However, in spite of that "gratitude," most women were not allowed to keep the jobs they had filled during the war when it was over. The few women who did remain in engineering after World War I must have been nearly invisible, because according to the *American Men of Science* directory for 1921, there were no women at all in engineering, although the 1920 Census reported 41 women out of a total of 130,000 engineers (Rossiter, 1992; Schneider & Schneider, 1993).

One of these 41 women was probably Elsie Eaves (Fig. 3.5), a 1920 civil engineering graduate from the University of Colorado who had just embarked on what would be an illustrious career. Eaves was instrumental in establishing a national

Fig. 3.5 Elsie Eaves (Photograph courtesy of Society of Women Engineers Records, Walter P. Reuther Library, Wayne State University)

organization for women engineers: the Society of Women Engineers and Architects. In 1919, she and several of her colleagues at the University of Colorado wrote to engineering schools across the country, asking for information on women engineering students and graduates. They found 63 women enrolled in engineering at 20 universities—including 43 at the University of Michigan alone! The Michigan women had organized a group of their own, the T-Square Society, in 1914. Most schools, however, did not yet admit women to their engineering programs. One professor responded to Eaves' letter: "I would state that we have not now, have never had, and do not expect to have in the near future, any women students registered in our engineering department" (Kindya, 1990). In the end, there were too few women engineers around the country, and they were too scattered geographically to keep the organization alive, so it folded (Kindya, 1990).

Women engineers made little headway in the 1920s and 1930s. Despite the economic boom in the 1920s and the accompanying optimism and expansion, women in the professions generally did not benefit from the strong economy. Out of 197 fellowships awarded in engineering during one period of the 1920s, one went to a woman. However, the number of women engineering graduates did grow by 113 from 1920 to 1929, reaching 158. And 53 engineering colleges (an increase from 35 in 1920) now admitted women (Turner, 1954).

Those women who were able to graduate with a degree and find employment still faced significant obstacles in advancing in their careers. After graduation, a man would start as a junior engineer and could generally expect, if his performance was satisfactory, to be promoted to senior engineer and then project engineer. Then, he would move on to management, with each step representing increased status, power, and salary. Women engineers, however, were usually limited to desk work, sometimes bordering on clerical work, and did not have access to plant or fieldwork, which denied them the necessary experience to move up the ladder. Many higher-level positions required travel to remote locations, behavior not acceptable for "gentle women." Women often needed advanced training—not generally required of men. Although a few women were able to hold positions with professional responsibility, many others operated on the fringes of the profession as writers, editors, secretaries, librarians, industrial teachers, or laboratory assistants—all more "acceptable" positions for women than being an actual engineer (Barker, 1994).

3.6 The Great Depression

Men and women alike were forced to cope when the stock market crashed in September 1929, and the Great Depression followed. Many women had to sacrifice personal ambition and accept a life of economic inactivity (LeBold & LeBold, 1998). Thus, many women engineers either voluntarily or involuntarily ceased their careers after 1929. Jobs in general were scarce, and what jobs were available were unlikely to be given to women.

However, as college enrollments dwindled, colleges began to look for women as students. Twenty-seven more engineering schools began to admit women, and the number of female engineering graduates increased by 156 during 1930–1939. With a new emphasis on graduate education for engineers as a result of technology developments during World War I, it is interesting that six schools that excluded women at the undergraduate level allowed them to pursue graduate engineering degrees. Prior to 1942, 18 women had received master's degrees in engineering across the entire USA. One woman did receive an engineering Ph.D. in 1920, but no more women earned that capstone degree again until three women did so in the 1930s. However, women still accounted for only about 3 out of every 1000 engineers. And no women were on the engineering faculty at any of the 20 largest doctoral universities in the country in 1938 (Barker, 1994; Turner, 1954).

By 1938, the percentage of women in engineering represented significantly less than one half of 1% of all engineers. Although it was estimated that about 1000 women engineers and architects were trained in the USA, the 1938 edition of *American Men of Science* reported eight women, representing 0.2% of its approximately 3500 engineers (Rossiter, 1992, 1995). The 1940 Census listed 730 women employed as engineers (see Table 3.1), less than 0.3% of the total; but many of those are thought not to have the education credentials (U.S. Government Printing Office, 1948). And in 1938, there were zero women engineers among the almost 20,000 engineers employed in the federal classified civil service (U.S. Government Printing Office, 1948).

The highly visible women of whom we know about today were either linked with a male relative who was making a significant impact in the engineering field or so outstanding on their own that their accomplishments could not be ignored. Female engineering students were still aliens in a man's world and had to deal with both implicit and explicit constraints in their educations and in their jobs, constraints which often relegated them to the margins of the profession. In addition, the number of women in engineering was so small that no feminized branches of engineering even had a chance to develop (Barker, 1994). Most women were still not admitted

Table 3.1 Distribution by branch of engineering of men and women employed as professional engineers, 1940

Branch of engineering	Total	Number		Percent		Percent of women in total
		Men	Women	Men	Women	
All employed professional engineers	245,288	244,558	730	100.0	100.0	0.3
Civil engineers	80,362	80,171	191	32.8	26.2	0.2
Mechanical engineers	82,443	82,255	188	33.6	25.8	0.2
Electrical engineers	53,267	53,103	164	21.7	22.5	0.3
Industrial engineers	9283	9209	74	3.8	10.1	0.8
Mining and metallurgical engineers	8813	8739	74	3.6	10.1	0.8
Chemical engineers	11,120	11,081	39	4.5	5.3	0.4

to the professional engineering societies in their field of expertise, and a nationwide organization for women engineers had not yet been established.

3.7 World War II

World War II, in similar fashion to World War I, presented opportunities for women to assist in the war effort. This movement of women back into the workforce is often personified by famed metalworker and poster woman "Rosie the Riveter." Rosie (Fig. 3.6) is portrayed as a powerful woman, pictured in a headscarf while displaying large arm muscles (her sleeves are rolled up) with the caption "We Can Do It!" (LeBold & LeBold, 1998; Rossiter, 1995). Although World War I had seen the serious use of technology for airplanes, submarines, tanks, bombs, and other equipment, technological development and organized scientific research in the ensuing 20 years were such that the machines of World War II were more destructive and more powerful. Consequently, they needed engineers and technicians to design,

Fig. 3.6 Rosie the Riveter (Courtesy National Archives and Records Administration)

build, and maintain them. Many of the men with the skills to fill these engineering and technical jobs were now in the military (Barker, 1994).

Women were encouraged to enter the workforce because the men were gone. Indeed, they began to be invited into the good positions, not just those invisible jobs discounted as "women's work" (such as silk hose manufacturing) (Barker, 1994). People were needed at drawing boards and in engineering shops to keep planes, tanks, and other essential materials rolling off the assembly line—everywhere that the war effort needed support. The problem for the government was that with the men gone, women and Blacks, the two reserve labor forces of the country, were the only ones available to fill these positions. However, there were not enough appropriately trained women or Blacks. In engineering, the shortages were so great that women were vigorously recruited during World War II (Rossiter, 1995; Bix, 1999).

The Office of War Information (OWI) and the War Manpower Commission began issuing literature and propaganda glorifying women as scientists and engineers to bolster the war effort. In 1942, a film, *Women in Defense*, narrated by Katharine Hepburn from a script written by Eleanor Roosevelt, urged women to go to work in government or on scientific projects. In 1942, the movie *Madame Curie*, starring award-winning actress Greer Garson, was released, further glorifying women's contributions to science. New training programs for scientists and engineers were established, and, by 1943, women (and, to some extent, Blacks) were being specially recruited and trained for jobs in industry. Bright high school students were sought out and urged to major in science in college through the Westinghouse National Science Talent Search (established in 1943) and the Bausch and Lomb Science Talent Search (established in 1944). Both programs had women among their early winners, and the Westinghouse program required that the percentage of women winners be proportional to female entrants. Books and articles were released during the 1942 through 1945 period urging women to pursue careers in science and engineering. Edna Yost's book *American Women of Science* (1943), which lauded women's past and current contributions to science, painted a bright picture for women in these fields (Rossiter, 1995).

To fill the personnel gap, the U.S. Government began running crash courses in science, engineering, and management for both men and women after the war began in 1941. Some specific courses were targeted at helping women become engineering aides and engineering cadettes. Many engineering schools set up special engineering training courses for women sponsored by the War Department, the Signal Corps, the Ordinance Department, and the Air Force. Twenty-nine institutions that had heretofore excluded women, including the Carnegie Institute of Technology, Columbia University's School of Engineering, and Rensselaer Polytechnic Institute, admitted women engineering students for the first time between 1940 and 1945 to train them in support of the war effort (Barker, 1994; LeBold & LeBold, 1998; Rossiter, 1995; U.S. Government Printing Office, 1948; Bix, 1999).

3.7.1 Engineering Aides, Engineering Cadettes, and Engineers

In 1942 and 1943, the call went out for more women engineers. Elsie Eaves, of the *Engineering News-Record*, warned women that the word "engineer" was being applied to two very different career paths: Those few women with degrees in the field could expect to be hired into professional positions. But those nonengineering college female graduates with a few additional courses in drafting and machine testing, and those women without college degrees, were hired into subprofessional jobs as "engineering aides" or "engineering cadettes." Sometimes, they were temporary assistants to men who had been promoted from lower positions within the organization. Nonetheless, there were plenty of jobs for both kinds of "engineers" at this point. Even the federal government changed its longtime policy and began to hire women engineers in 1942 (Rossiter, 1995).

The demand was so great for the subordinate type of personnel (those classified as "aides" or "cadettes") that a special program was set up by the federal government as early as October 1940 in anticipation of entering World War II. The U.S. Office of Education administered this "defense training" program that was funded by a special appropriation. It was one of the first federal government efforts to increase and train scientific manpower. Manpower shortages were expected to be severe if the USA entered the war—the aircraft industry on Long Island and in New Jersey alone would require about half of all the engineering graduates nationwide. The first Emergency Defense Training (EDT) program course was offered in December 1940, and by June 1941, over 100,000 people (almost all men) had received training in engineering subjects (Barker, 1994; Rossiter, 1995).

At first, the Engineering, Science, and Management War Training (ESMWT) program offered training only at 4-year technical schools. Colleges were responsible for determining local training needs and developing courses to meet those needs, but not to provide a complete engineering education nor necessarily any academic credit for the courses (Barker, 1994). As the war continued and personnel shortages intensified, the program and the corresponding training were extended to a total of 227 colleges and universities, including several women's colleges, where courses were taught in elementary engineering, mathematics, chemistry, physics, and safety engineering. The ESMWT opened doors of opportunity to many women and some minorities (Rossiter, 1995). In a report issued in December 1942, the ESMWT reported classes offered at Clarkson, Cornell, the University of Rochester, Syracuse University, Rensselaer Polytechnic Institute, and Union College. Union College had, in fact, given courses to 145 women out of 817 students, including a special course in electricity and mechanics for "girl high school graduates employed on testing work in the Engineering Department of the General Electric Company" (Barker, 1994).

In 1943 and 1944, 21.8% of the enrollees were women and minorities. By 1945, over 1.8 million people had been trained including 280,000 women (15.7%) and 25,000 Blacks (1.4%) (Rossiter, 1995). About one-fifth of these women were enrolled in engineering drawing, while significant numbers were learning aeronautical engineering, inspection and testing, mapping and surveying, and engineering fundamentals; the rest were registered for other scientific and management courses (Barker, 1994).

Other departments of the federal government conducted their own programs to recruit and train women for technical positions. The Office of the Chief of Ordnance trained women with high school mathematics for civil service appointments as junior engineering aides during a 3-month intensive course at the University of Michigan. Female engineering aides for the Frankford and Picatinny Arsenals were trained at similar programs sponsored at Rutgers University, Drexel Institute, Temple University, and the University of Pennsylvania. The Signal Corps and the Army Air Forces provided technical training to hundreds of women who were often then assigned clerical duties, which resulted in a high turnover rate (Barker, 1994).

Some companies had other training requirements or could not wait for the results of the government programs to bear fruit. General Electric recruited women with degrees in math or physics and then gave them on-the-job training so that they could handle computations in GE's machine tool department. However, so few women existed with these types of credentials that the companies began reaching down further pursuing women still in college (Bix, 1999).

Aircraft companies started special programs for engineering aides and engineering cadettes. The Goodyear Aircraft Corporation developed a 6-month program in aeronautical engineering at the University of Cincinnati to prepare women as "junior engineers" (Barker, 1994).

The Vega Aircraft Corporation needed aircraft design engineers starting in 1941. In cooperation with Lockheed Aircraft Corporation and the California Institute of Technology, their engineering shortage was temporarily solved by hiring engineers trained in other disciplines—civil, mechanical, or electrical—from other industries and training them for aircraft work with 8 weeks of college-level instruction at Caltech, followed by 8 weeks of an on-the-job apprenticeship. Other technical and clerical employees with some experience took evening classes specially arranged with the University of California in aircraft engineering-related subjects. But soon, all available candidates for these programs had been trained. Next, Vega established a full-time training program at the University of California at Los Angeles (UCLA) to teach drafting for women. These women were not already employees but were guaranteed jobs if they finished the course. By 1943, Vega established a full-time 52-week program at Caltech for two groups of 20 full-time employees. The 40 students, all women, ranged in age from 18 to 49 years old and already worked in the engineering department at Vega. They were able to take a full college course in engineering, stripped of all nonessentials, and earn the chance to be upgraded to higher-level jobs (Barker, 1994).

In 1942, the Curtiss-Wright airplane company developed a plan for training "CW cadettes," young women with at least 2 years of college education, including 1 year of college-level mathematics. Seven colleges, some of which had never previously enrolled women—Cornell, Penn State, Purdue, Minnesota, Texas, Rensselaer Polytechnic Institute, and Iowa State—gave over 600 women a 10-month crash course of engineering, math, job terminology, aircraft drawing, engineering mechanics, airplane materials, theory of flight, and aircraft production. After their training, these cadettes were assigned to airplane plants to work in research, testing, and production (Barker, 1994; LeBold & LeBold, 1998; Bix, 1999).

The Engineering Cadette Program in the School of Electrical Engineering at Purdue University sponsored by Radio Corporation of America (RCA) helped convince skeptical faculty that women could excel technically and academically. Cadettes accounted for 20% of the engineering staff in some RCA plants, allowing women to move beyond being tokens (Barker, 1994).

By 1945, the number of women engineers listed on the National Roster had increased from 144 in 1941 to 395 (an amazing increase of almost 300% in 4 years), mostly at the bachelor's level. And many more engineering colleges and universities now at least had seen women as students (Rossiter, 1995). The number of women enrolled as engineering undergraduates had increased to 1800 by 1945 (Barker, 1994).

3.7.2 *Faculty*

Women were also encouraged to become faculty members at colleges and universities, temporarily replacing men who had been called into government service or into the military. In fact, a significant number of female scientists, including mathematicians and chemists, who were not working on other war projects, moved into the ranks of college and university faculty. In 1942, 2412 women scientists (including 50 women in engineering out of a total of 5394 engineering teachers) were on science faculties. By 1946, the number had increased to 7746 (a more than 200% increase in 4 years), and the number of women engineering faculty, including some women who taught at all-male institutions, had increased to 53 (Barker, 1994; Rossiter, 1995).

These women cadettes, engineering aides, engineers in industry, and faculty all made significant contributions to the war effort and to the advancement and acceptance of women in technical fields. All in all, about 45,000 women were trained for engineering jobs during World War II (Barker, 1994).

3.8 After World War II

When the war ended, millions of veterans—primarily men—came home expecting to be gainfully employed. Many of the women who had been employed found themselves no longer welcome in the workforce. In fact, these women were now expected to go home and raise babies or "go back to the kitchen" and not steal work from returning GIs. The special educational programs were shut down. And because many engineering schools still did not admit women, educational opportunities for women who wanted to pursue an engineering career were once again limited. By

January 1946, four million women had left the labor force (Barker, 1994; LeBold & LeBold, 1998).

By the time World War II ended, however, women were not quite as eager to accept the inevitable as they were when World War I ended. They began generating statistics and scientific data to prove that they were just as capable as men and were being treated unfairly. Still, even for those who managed to hold on to their positions, they were unable to advance as men did. Women were still channeled into less challenging work, kept in lower ranks, and paid lower salaries by employers who minimized their contributions (Rossiter, 1995).

A very significant impact of the war's end was the GI Bill, which provided funds for veterans' education. Schools that had been underutilized during the war were suddenly faced with a deluge of male students. It is estimated that 7.8 million veterans chose to take advantage of the GI Bill's educational provisions. Female enrollment at some coeducational schools had reached 50% or higher during the war, but as the demand for slots increased, many institutions introduced or reintroduced maximum quotas on female enrollment. Some colleges refused to accept applications from out-of-state women. Women were told that space was not available and that they would have to attend other institutions. For a few years, women who were already in the educational pipeline, women veterans, and wartime women "engineering aides" getting engineering undergraduate degrees contributed to a brief expansion in the number of women getting engineering degrees. Through the 1950s, the percentage of women dropped to just 25–35% of the total college undergraduate enrollment. The number of women graduating with engineering degrees declined as well, as shown in Table 3.2 (Rossiter, 1995).

Discrimination against women was widespread; women were systematically pushed out of science and engineering at the undergraduate level, at the graduate level, and as faculty. Men even took over women's dormitories, and many universities added temporary housing to accommodate the returning male GIs. Male veterans replaced female staff and faculty as well as female engineering students. There were no discrimination laws in place (these would not be enacted until the 1960s or later) to prevent any of this behavior. The nation was also extremely grateful to its veterans and felt that there was, in many cases, a duty to make their adjustment to civilian life as easy as possible. Objections or protests were few and far between and not effective. In addition, society as a whole was still quite ambivalent about the proper role for women. These ambitious women in engineering, who were certainly atypical, threatened presumed male and female spheres (Rossiter, 1995).

Table 3.2 Women earning undergraduate engineering degrees 1948–1953

Year	Number of degrees
1947–1948	191
1949–1950	171
1951–1952	60
1952–1953	37

The *Outlook for Women in Architecture and Engineering*, a bulletin of the Women's Bureau of the U.S. Department of Labor published in 1948, painted a bleak picture for women in engineering (U.S. Government Printing Office, 1948):

> *Advancement for women in engineering is conceded to be difficult. They seldom follow the usual line from junior engineer to senior engineer to project engineer, nor are they often transferred to nonengineering work in sales, purchasing or administration. Usually limited by custom to office work, as compared with field or plant work, women engineers rarely find opportunities to obtain the rounded experience necessary for normal progression. The fact that many of the positions representing advancement often require field work or travel to remote locations further reduces their chances. However, a few women have broken through these bounds.*

Women engineers, who had been welcomed with open arms during the war years, were now told to abandon the idea of an engineering career. The Women's Bureau reported that there were 950 women in engineering as of 1946–1947. These women constituted only 0.3% of the total 317,000 engineers in the USA. Of course, this was much higher than the eight women reported to be practicing engineering in the 1949 edition of *American Men of Science*, which, despite its title, reported on all individuals practicing in the sciences. The U.S. Census of 1950 counted 6475 women employed as engineers, of whom only 41% had 4 years or more of college and 17% had not graduated from high school. Probably about 3600 of these women were truly engineering professionals. There were now an adequate number of women in engineering to finally constitute enough of a critical mass to move the cause of women in engineering forward, both to seek more women as engineers and to improve career opportunities for those already in the field (Barker, 1994; Rossiter, 1995; Kindya, 1990; U.S. Government Printing Office, 1954).

In fact, women seeking careers or advancement in their existing careers and students studying engineering started to band together for mutual support and job-hunting help. In the late 1940s, the war-support effort and the associated defense industry had made cutbacks across the board. Employment for many engineering disciplines was at a low. Many professional societies still excluded women, either outright or by refusing them full membership, and thus had very few women among their membership as shown in Table 3.3. But this time, efforts for mutual support among women engineers were enabled by a nationwide infrastructure that had not existed in 1919. Now there was the telephone, the automobile, the superhighway, and the railroad (Kindya, 1990).

Table 3.3 Membership in the Founder societies, 1946 (ASCE, 2000a)

Organization	Total membership	Number of women members
American Institute of Electrical Engineers	24,526	14
American Society of Civil Engineers	21,100	23
American Society of Mechanical Engineers	20,060	33
American Institute of Mining and Metallurgical Engineers	12,600	26
American Institute of Chemical Engineers	5788	5

Female engineers fueled by isolation and discrimination, and the national infrastructure, founded the Society of Women Engineers (SWE) in 1949–1950 (LeBold & LeBold, 1998).

3.9 Society of Women Engineers

SWE traces its founding back to a group of female students at Drexel University in Philadelphia and a gathering of Cooper Union and City College of New York graduates and students in New York City. The Drexel students had begun meeting in 1948, calling themselves the "Philadelphia District of the Society of Women Engineers." The New York Group had been meeting informally for several years and had even had meetings with the Drexel students. In November 1949, the New York women officially started the New York Organization with a designated treasurer and bylaws chair. Within a few months, the New York Organization would elect a president as well. Women from Boston and Washington, DC, also considered themselves part of the "National Organization" (Rossiter, 1995; Kindya, 1990; SWE, 2015a, Hayes, 2000).

The Philadelphia District's newsletter announced in November 1949 that the National Organization would hold a convention in the spring or fall of 1950. The meeting at Green Engineering Camp of The Cooper Union was held on May 25, 1950, and more than 60 women attended from the four districts of Philadelphia (including the Drexel students), New York, Boston, and Washington, DC. The women who attended the Camp Green meeting represented two major constituencies: (1) the pioneers, those who had been trained and were working as engineers prior to or during World War II, and (2) the engineering aides who had returned to college after the war to get their engineering degrees. Many issues were decided at this meeting including whether the organization would be comprised of just graduate engineers or would also include student members. That meeting is generally referred to as the official founding of SWE (Rossiter, 1995; Kindya, 1990; SWE, 2015a, Hayes, 2000). The first president, Beatrice Hicks (Fig. 3.7), was elected; membership requirements were established; and dues were set. The initial board of directors consisted of five officers, three directors-at-large and one director from each section. A nominating committee was established (Rossiter, 1995; Kindya, 1990; SWE, 2015a, Hayes, 2000).

During the rest of 1950, the organizational structure of SWE was refined. Nominating, membership and finance, and convention committees were established. The logo developed by the Drexel students was refined by the emblem committee, and a publications committee was established that included Elsie Eaves on its editorial board. SWE would put together an official publication; it would be titled *Journal of the Society of Women Engineers*, and it would be released quarterly. It was also decided that SWE would not participate in controversial political matters and that nonprofit status would be sought (Hayes, 2000).

Fig. 3.7 Beatrice Hicks (Photograph courtesy of Society of Women Engineers Records, Walter P. Reuther Library, Wayne State University)

The first issue of the *Journal* was published in the spring of 1951. That issue reported that Lillian Moller Gilbreth, who was a staunch supporter of SWE until her death in 1972, had accepted honorary membership. Although technical articles were published in the first few issues, by mid-1952, the *Journal* turned its focus to reports of section activities, publicity received by the sections, notes from members, and listings of new members. Not until the 1980s would SWE's publication (renamed *U.S. Woman Engineer* by then) feature technical articles again (Hayes, 2000).

In 1951, the SWE membership approved the organization's bylaws, and the Pittsburgh Section was chartered. In addition, a convention was held in New York City. SWE was incorporated in February of 1952 and registered as a corporation in the District of Columbia. Significantly, in 1952, SWE was invited to participate in the "Centennial of Engineering" to celebrate 100 years of the ASCE. This lent credibility to the organization since all the major professional societies would be represented. The Chicago and Detroit Sections were chartered in 1952, and the organization became truly national in 1953 when the Los Angeles Section was chartered (Hayes, 2000).

As contained in the Articles of Incorporation, the objectives of the organization were:

> *To inform the public of the availability of qualified women for engineering positions; to foster a favorable attitude in industry towards women engineers; and to contribute to their professional advancement.*
>
> *To encourage young women with suitable aptitudes and interest to enter the engineering profession, and to guide them in their educational programs.*
>
> *To encourage membership in other technical and professional engineering societies, participation in their activities, and adherence to their code of ethics* (Kindya, 1990; Hayes, 2000).

By 1952, SWE had established the Achievement Award and recognized its first recipient. This award, the highest tribute given by SWE, honors a woman engineer who has made outstanding contributions in a field of engineering over a significant period of time. The first few recipients were (Kindya, 1990; Hayes, 2000, SWE 2015b):

1952: Mária Telkes, "In recognition of her meritorious contributions to the utilization of solar energy" (Fig. 3.8)

1953: Elsie Gregory MacGill, "In recognition of her meritorious contributions to aeronautical engineering"

Fig. 3.8 Maria Telkes (Photograph courtesy of Society of Women Engineers Records, Walter P. Reuther Library, Wayne State University)

1954: Edith Clarke (Fig. 3.9), "In recognition of her many original contributions to stability theory and circuit analysis"
1955: Margaret H. Hutchinson, "In recognition of her significant contributions to the field of chemical engineering"
1956: E'lise F. Harmon, "In recognition of her significant contributions to the area of component and circuit miniaturization"
1957: Rebecca H. Sparling, "In recognition of her meritorious contributions to high temperature metallurgy and non-destructive testing of metals"

Finally, there was an organization willing to accept women engineers, recognize them for their accomplishments, and encourage other women to be engineers.

The organization grew rapidly, numbering 350 in 1953 and surpassing 500 by 1958. The group received decent publicity too—early activities of SWE were reported in *The New York Times* (Rossiter, 1995; New York Times, 1951a, b, c, 1952;). A Professional Guidance and Education committee was established to provide information that members could use for outreach to high schools and PTA programs. The committee printed a booklet containing speeches that had been

Fig. 3.9 Edith Clarke (Photograph courtesy of Society of Women Engineers Records, Walter P. Reuther Library, Wayne State University)

delivered at the Centennial of Engineering by SWE members including a paper from Margaret Ingels titled “Petticoats and Slide Rules.” SWE participated in the 75th Anniversary of the ASME at Stevens Institute of Technology and established a statistics committee. In 1956, a headquarters fund was started to provide income for the eventual establishment of a Headquarters office, the scholarship fund was established, and a Board of Trustees was elected to manage both funds (Hayes, 2000).

In 1957, SWE was granted permission to office in the same building as the established professional engineering societies—the Engineering Societies Building—in New York City. Later, when those societies moved to the United Engineering Center—across from the United Nations in Manhattan—SWE moved with them. SWE archives were established as well (Hayes, 2000).

Beginning in 1958, the interests of engineering undergraduate students were seriously addressed, and the first SWE scholarship was announced. It was called the Lillian Moller Gilbreth Scholarship in honor of SWE’s staunch supporter and first honorary member (Hayes, 2000).

The 1960s saw much growth and many activities. By 1961, SWE was in its new quarters in the United Engineering Center, and because of the launch of Sputnik, the USA wanted engineers in government and industry—even if they were women! Membership had reached over 1000. More engineering schools began to accept women students. SWE hosted the first International Conference of Women Engineers and Scientists at the United Engineering Center. SWE became an active liaison with the ECPD and an associate member of the Engineers Joint Council (Hayes, 2000).

Growth in sections, student membership, activities, and the organization itself has continued unabated since then. The Fellow grade was established in 1979 as a way of honoring long-standing and hard-working members who have paved the way for SWE. In 1982, SWE had 49 professional sections, 194 student sections, and almost 13,000 members. SWE’s archives have been housed at the Reuther Library at Wayne State University in Detroit, Michigan, since 1993 (Hayes, 2000).

In 2024, SWE’s membership has risen to more than 40,000 individual members in more than 100 professional sections and 300 collegiate sections in the USA. SWE is undertaking efforts internationally as well—in Africa, Brazil, Europe, and India (SWE, 2024).

Women engineers have come a long way, but the goal of gender parity (50/50) remains elusive. The barriers seem to have fallen in many areas—particularly in educational institutions and professional societies. Women can have fulfilling engineering careers in which they are paid equally to the men who are doing the same work. The USA—and the world—just needs more of them!

Author’s Note Adapted in large part from the first three chapters of *Engineering Women: Re-visioning Women’s Scientific Achievements and Impacts* by Jill S. Tietjen, P.E., the first volume in the Springer Women in Engineering and Science series.

References

AIChE. (1999). A brief history of AIChE. Retrieved from www.aiche.org/welcome/history.htm. Accessed 20 Aug 1999

AIChE. (2024). *Institute Milestones, AIChE.* Retrieved March 31, 2024, from https://www.aiche.org/about/institute-milestones

Ambrose, S., Dunkle, K., Lazarus, B., Nair, I., & Harkus, D. (1997). *Journeys of women in science and engineering: no universal constants*. Temple University Press.

ASCE. (2000a). ASCE profile. Retrieved from www.asce.org/aboutasce/profile.html. Accessed 15 Dec 2000

ASCE. (2000b). 150 years of civil engineering. Retrieved from www.asce.org/150/1506years.html. Accessed 15 Dec 2000

ASCE's First Woman President Assumes Office Saturday. (2003). Wastewater digest. Retrieved from https://www.wwdmag.com/water/news/10908068/asce039s-first-woman-president-assumes-office-saturday. Accessed Nov 2003

ASME. (2000). The History of ASME International. Retrieved from www.asme.org/history/asme-hist.html. Accessed 15 Dec 2000

ASME's First Female President Honored by Alma Mater. (2014). Retrieved from https://www.asme.org/topics-resources/society-news/asme-news/asmes-first-female-president-honored-alma-mater. Accessed 29 May 2014

ASSE. (2000). *Society history*. Retrieved December 19, 2000, from www.asse.org/hsoci.htm

Barker, A. M. (1994). "Women in Engineering During World War II: A Taste of Victory." Unpublished

Bix, A. S. (1999). 'Engineeresses' 'Invade' Campus: Four Decades of Debate over Technical Coeducation. In *Proceedings of the 1999 International Symposium on Technology and Society—Women and Technology: Historical, Societal, and Professional Perspectives, July 29–31, 1999, New Brunswick, NJ* (pp. 195–201).

Founder Societies of the United Engineering Foundation (2000). Retrieved from www.uefoundation.org/fndsoc.html. Accessed 15 Dec 2000

Goff, A. C. (1946). *Women can be engineers*. Edwards Brothers, Inc.

Hayes, D. W. (2000). *Five decades of the society of women engineers*. Society of Women Engineers.

IDIS. (2000a). Historical highlights. Retrieved from www.idis.com/aime/history.htm. Accessed 15 Dec 2000

IDIS. (2000b). WAAIME. Retrieved from www.idis.com/aime/WAAIME.HTM. Accessed 19 Dec 2000

IEEE. (2000a). The origins of IEEE. Retrieved from www.ieee.org/organizations/.ical_articles/history_of_ieee.html. Accessed 15 Dec 2000

IEEE. (2000b*). Welcome to the IEEE Fellow Program: Our history.* Retrieved December 19, 2000, from www.ieee.org/about/awards/fellows/fellows.htm

IEEE. (2024). Computer society. Martha Sloan. Retrieved from https://www.computer.org/profiles/martha-sloan. Accessed 31 March 2024

IIENet. (1999a). Retrieved August 20, 1999, from www.iienet.org/Aboutg.htm, p. 1.

IIENet. (1999b). Retrieved August 20, 1999, from www.iienet.org/historg.htm

Ingels, M. (1952). *Petticoats and slide rules*. Western Society of Engineers. September 4, 1952, and later published in *Midwest Engineer*.

Kass-Simon, G., & Farnes, P. (1990). *Women of science: Righting the record.* Indiana University Press.

Kindya, M. N. (1990). *Four decades of the society of women engineers*. Society of Women Engineers.

LeBold, W. K., & LeBold, D. J. (1998). Women engineers: a historical perspective. *ASEE Prism, 7*(7), 30–32.

National Academy of Engineering. About the NAE. (2000). Retrieved from www.nae.edu/nae/naehome.nsf/weblinks/NAEW-4NHMQM?opendocument. Accessed 19 Dec 2000

New York Times. (March 11, 1951a). Women engineers see field widening. p. 47.
New York Times. (March 12, 1951b). U.S. Agencies Seek Women Engineers. p. 19.
New York Times. (August 6, 1951c). Again heads women engineers. p. 14.
New York Times. (March 16, 1952). Women could fill engineering jobs: Trade society, meeting here, told they represent untapped 'source of qualified talent. p. 79.
Read, P. J., & Witlieb, B. L. (1992). *The book of women's firsts*. Random House.
Rossiter, M. W. (1992). *Women scientists in America: Struggles and strategies to 1940*. The Johns Hopkins University Press.
Rossiter, M. W. (1995). *Women scientists in America: Before affirmative action 1940–1972*. The Johns Hopkins University Press.
Schneider, D., & Schneider, C. F. (1993). *The ABC-CLIO companion to women in the workplace*. ABC-CLIO.
SME Past Presidents. (2024). Retrieved from https://www.smenet.org/What-We-Do/SME-Past-Presidents. Accessed 31 March 2024
Society for Mining, Metallurgy and Exploration. (2015). The Woman's Auxiliary to the American Institute of Mining, Metallurgical, and Petroleum Engineers, Inc. a division of the Society for Mining, Metallurgy and Exploration (SME). Retrieved from http://www.smenet.org/waaime/. Accessed 5 Apr 2015
SWE. (1959). Elsie Eaves—Pioneer from the West. *SWE Newsletter*, p. 3.
SWE. (1999). Retrieved from www.swe.org/SWE/Awards/achieve3.htm. Accessed 1 Sept 1999
SWE. (2015a). The SWE Story. Retrieved from http://societyofwomenengineers.swe.org/membership/history/2454-the-swe-story. Accessed 24 May 2015
SWE. (2015b). Individual awards overview: Achievement award. Retrieved from http://societyofwomenengineers.swe.org/awards/individual-awards. Accessed 24 May 2015
SWE. (2024). *About SWE, Society of Women Engineers*. Retrieved April 2, 2024, from https://swe.org/about-swe/
Tau Beta Pi. (1999). *Integrity and Excellence in Engineering*. Retrieved August 20, 1999, from www.tbp.org/TBP/INFORMATION/Info_book/membership
The Phi Beta Kappa Society: A Short History of Phi Beta Kappa. (2000). Retrieved from www.pbk.org/history.htm. Accessed 19 Dec 2000
Turner, E. M. (1954). Education of women for engineering in the United States 1885–1952, Dissertation, New York University. Ann Arbor, MI: UMI Dissertation Services.
U.S. Government Printing Office. (1948). *The outlook for women in architecture and engineering* (pp. 5–33). Bulletin of the Women's Bureau No. 223-5.
U.S. Government Printing Office. (1954). *Employment opportunities for women in professional engineering*. Women's Bureau Bulletin No. 254.
United Engineering Foundation (2000a). *An overview of the United Engineering Foundation: History. Retrieved* December 15, 2000, *from* www.uefoundation.org/overview.html
United Engineering Foundation (2000b). *Founder Societies of the United Engineering Foundation.* Retrieved from December 15, 2000, www.uefoundation.org/fndsoc.html
University of Arizona. (2023). College of engineering news, 1978 mining alumna in the business of ridding industry barriers. Retrieved from https://news.engineering.arizona.edu/news/1978-mining-alumna-business-ridding-industry-barriers. Accessed 28 Nov 2023
WAAIME. (2000). Retrieved from www.idis.com/aime/WAAIME.HTM. Accessed 19 Dec 2000
Wastewater Digest. (2003). ASCE's first woman president assumes office Saturday. Retrieved from https://www.wwdmag.com/water/news/10908068/asce039s-first-woman-president-assumes-office-saturday. Accessed 14 Nov 2003
Welcome to the IEEE Fellow Program: Our History. (2000). Retrieved from www.ieee.org/about/awards/fellows/fellows.htm. Accessed 9 Dec 2000

Chapter 4
History of and Trends in Women's Participation in Academic Computing

Kaitlyn N. Stormes, Kathleen J. Lehman, Elizabeth Litzler, and Lecia Barker

4.1 History of and Trends in Women's Participation in Academic Computing

Women have been central to academic computing since the field's foundation. Indeed, many of the earliest and most important voices in computing have been women in academia. Ada Lovelace was a mathematician who developed the first computing program in the 1840s. Grace Hopper, who developed the first compiler, held a PhD in mathematics and mathematical physics from Yale University and was a professor of mathematics at Vassar College in the 1930s before she began her storied career in the Navy. Furthermore, Gertrude Blanch held a PhD in mathematics from Cornell University and taught at Hunter College before assuming her role as the director of mathematics for the Computing Tables Project, a sophisticated computing organization that was the predecessor to the National Institute of Standards and Technology. During World War II and the postwar period of the 1950s, women with strong academic backgrounds in mathematics served as "coders" and "human computers" (Ensmenger, 2012). The inception of computers saw

K. N. Stormes (✉)
University of Georgia, Athens, GA, USA
e-mail: kaitlyn.stormes@uga.edu

K. J. Lehman
University of South Carolina, Columbia, SC, USA
e-mail: kl86@mailbox.sc.edu

E. Litzler
University of Washington, Seattle, WA, USA
e-mail: elitzler@uw.edu

L. Barker
University of Colorado Boulder, Boulder, CO, USA
e-mail: lecia.barker@colorado.edu

J. P. Carpenter (ed.), *Women in Academia*, Women in Engineering and Science,
https://doi.org/10.1007/978-3-031-97938-5_4

Women of Color[1] undertaking many early computer programming roles for private companies, NASA, and the military (Ensmenger, 2012). Following the war's conclusion and the return of men to the workforce, the computing field experienced a surge and was increasingly perceived as a profession best suited to men and, as such, began to evolve to exclude women (Ensmenger, 2012). This shift is reflected in the changing interest in college students' interest in the field: in the 1970s, undergraduate women kept pace with their male peers' interest in pursuing a computer science major (Eagan et al., 2017; Stolzenberg et al., 2019, 2020; Stolzenberg & Eagan, 2018). However, in the 1980s, the gender gap began to widen and has continued to grow.

As we will demonstrate in this chapter, these shifts in undergraduate student interest mirror shifts in women's participation in academic computing at the master's, doctoral, and faculty levels. As described below, all women—especially Women of Color—are sorely underrepresented at all levels of academic computing (see Table 4.1), yet they have continued to make important contributions to the field. We use the term "academic computing" to describe individuals doing research, scholarship, and teaching in computing within the confines of academic institutions and organizations (e.g., colleges, universities, national labs, research organizations). In this chapter, we aim to trace women's waxing and waning participation in academic computing since the 1960s (as far back as national data on this topic exists) and discuss the role of key organizations in promoting the success of women in academic computing. Throughout the chapter, we seek to elevate the intersectional experiences of Women of Color wherever possible, as they may experience multiple layers of marginalization via sexism (as women) and/or racism (as People of Color; Ong et al., 2011; Rodriguez et al., 2020).

4.2 Trends in Women's Computing Degree Attainment

Historical trends from 1964 to as recent as 2021 portray perpetual gender and racial/ethnic equity gaps across all levels of degrees earned in computing, with some narrowing of these achievement gaps seen for all degree levels in the early 1980s and some widening that occurred again in the early 2000s specifically for undergraduates (Sax et al., 2017). Today, women represent 58% of all undergraduates, 62% of all master's recipients, 56% of all doctoral degree recipients, and 44% of all college tenure-track faculty. Yet, they comprise only 22% of bachelor's degrees, 34% of master's degrees, 25% of doctoral degrees, and 14% of tenure-track faculty in computing (Fast Facts: Women Working in Academia, 2020). While the need to increase

[1] We rely on the 2022 National Academies of Sciences, Engineering, and Medicine "Transforming Trajectories for Women of Color in Tech" report for our definition of Women of Color, which includes women who identify as American Indian or Alaska Native, Asian, Black or African American, Native Hawaiian/Pacific Islander, and Hispanic and those who are multiracial not Hispanic.

Table 4.1 2020–2021 National Data on Women's Degree Attainment in Computer and Information Sciences and Support Services by Academic Degree Level and Race/Ethnicity

	Bachelor's				Master's				PhD			
	All fields[a]		Computing[b]		All fields[c]		Computing[d]		All fields[e]		Computing[f]	
	#	%	#	%	#	%	#	%	#	%	#	%
Total students	2,066,445	100	104,874	100	866,894	100	104,874	100	194,059	100	2572	100
Total women	1,205,681	58	22,953	22	536,194	62	22,953	22	108,692	56	650	25
White women	673,095	33	8429	8	293,842	34	8429	8	60,991	31	172	7
Black women	134,435	7	2544	2	69,104	8	2544	2	11,522	6	51	2
Hispanic women	201,592	10	2483	2	62,041	7	2483	2	9808	5	12	<1
Asian/Pacific Islander women	93,557	5	5310	5	34,327	4	5310	5	13,117	7	52	2
American Indian/ Alaska Native women	6138	<1	71	<1	2648	<1	71	<1	410	<1	0	0
Women from two or more races	48,366	2	1049	1	15,094	2	1049	1	3479	2	11	<1
Nonresident women	48,498	2	3067	3	59,138	7	3067	3	9365	5	352	14%

Notes: These data were derived from the following sources:
[a] Data for total bachelor's across all fields were derived from NCES Table 322.20
[b] Data on the total students who earned degrees in computing came from NCES Table 325.35, but data for women in computing came from NCES Table 322.50
[c] Data for the total master's across all fields were derived from NCES Table 322.20
[d] Data on the total students who earned degrees in computing came from NCES Table 325.35, but data for women in computing came from NCES Table 322.50
[e] Data for the total PhD across all fields were derived from NCES Table 324.20
[f] Data on the total students who earned degrees in computing came from NCES Table 325.35, but data for women in computing came from NCES Table 324.35

the representation of all women is clear from the gender gaps described above, there is an even greater need to broaden participation for women from systemically minoritized racial/ethnic backgrounds, such as those from Black, Latina, and Indigenous communities as these individuals tend to be among those least represented across all levels of academic computing.

Considering the historical context of the computing field mentioned above, the current gender and racial/ethnic equity gaps across all levels of degree attainment in computing are unsurprising, but how did we get here? It is well documented that these equity gaps exist at all levels of academic and professional computing due to unwelcoming, hostile, sexist, and racist computing climates (Beyer, 2014; Lane et al., 2021; Lunn et al., 2021), in addition to other workplace issues, including bias in hiring practices and promotion processes (Ashcraft et al., 2016; Whitney, 2021). To address this question, the following sections trace women's participation in academic computing across all levels, beginning with undergraduate degrees through faculty positions.

While there are a few possible data sources from which these data can be derived, we rely on data from the National Center for Education Statistics (NCES) and Computing Research Association's (CRA) Taulbee Survey reports. The NCES data was selected to portray general trends across all academic disciplines, including computing (where possible). The Taulbee Survey provides an annual overview of students and faculty in computing departments at PhD-granting institutions (Zweben & Bizot, 2023), which is useful in providing us with a more current snapshot of the state of the field. Furthermore, it is the only data source to our knowledge that has information for postsecondary students *and* faculty of varying ranks in computing by gender and race/ethnicity. Thus, we use this source consistently throughout the chapter as it (a) covers all levels of academic computing and (b) provides the closest comparison to national data for undergraduate and graduate student education in computing. However, we acknowledge that the Taulbee Survey data have limitations, namely that given its focus on PhD-granting institutions, it does not provide complete information on computing students or faculty at other institutional types.

4.3 Undergraduate Students in Computing

Notably, existing literature highlights that the more gender diverse an institution's student body is, the more diverse the faculty is (Main et al., 2020). Further, in exploring the correlation between student and faculty demographics, scholars such as Main et al. (2020) found that departments that award more bachelor's degrees to women from African American/Black, Hispanic/Latina, and Asian/Asian American backgrounds are more likely to employ relatively more women faculty of the same race. Importantly, the composition of advanced degree student bodies relies on eligibility criteria linked to undergraduate degrees in computing, emphasizing the need to address disparities at the undergraduate level.

Over approximately the last six decades, the gender gap in computing bachelor's degree attainment has varied greatly. For instance, in 1964, women earned only 5% of degrees awarded in computing. This percentage increased sevenfold, such that by

1983, women earned 37% of all computing bachelor's degrees. However, as stated above, more recent degree attainment data highlights that as of 2020, women earned only 22% of computing degrees awarded at the baccalaureate level (see Fig. 4.1). Further, Fig. 4.2 illustrates that white women and Asian women have consistently earned the majority of computing bachelor's degrees awarded to women over the last several years. The next largest racial/ethnic group of women earning bachelor's degrees in computing is Hispanic/Latina women—a trend consistent with bachelor's degree attainment rates for women across all fields (see NCES Table 322.20).

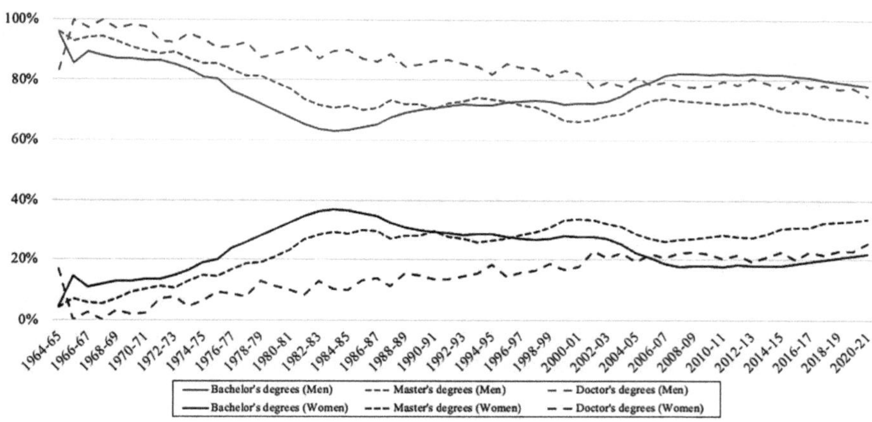

Fig. 4.1 Degrees in computer and information sciences conferred by postsecondary institutions, by level of degree and sex of student: academic years 1964–1965 through 2020–2021. Note: These data were derived from the National Center for Education Statistics (NCES) Table 325.35

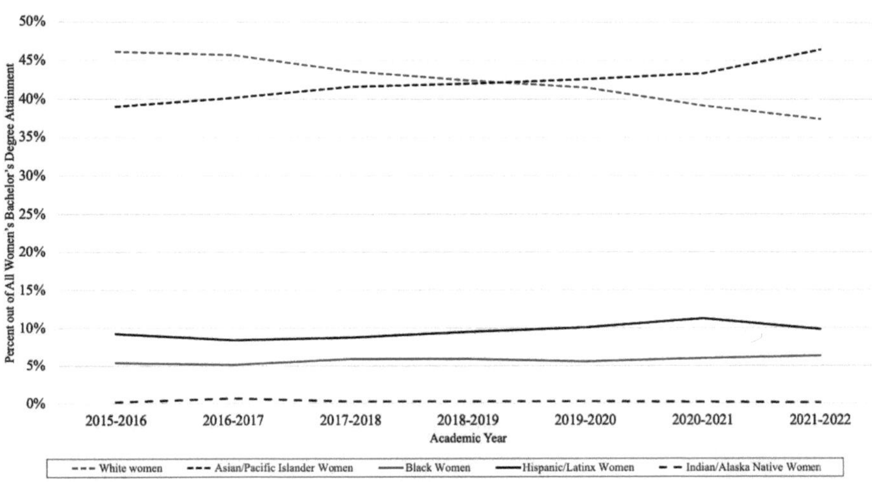

Fig. 4.2 Bachelor's degrees awarded to women in computing by race/ethnicity: academic years 2015–2016 through 2021–2022. Note: This visual portrays trends from the last 7 years of available data on the percentage of bachelor's degrees awarded to residential women by race/ethnicity. Data for this visual came from Table B7 in respective CRA Taulbee Reports from 2015 to 2021

4.4 Graduate Students in Computing

In acknowledging that women likely need to have earned a PhD in a computing discipline for them to assume academic positions, we now pivot to trace the gender and racial/ethnic equity gaps for women in computing *graduate* education, beginning at the master's degree level before transitioning to the representation of women earning PhDs in computing.

Master's Degree Level A review of data from the last 60 years indicates that the gender gap for those earning master's degrees in computing has fluctuated greatly. As shown in Fig. 4.1, only 4% of women earned master's degrees in computing in 1964, but in 2000, as many as 34% of women did. With some variation since then, as of 2020, women again earned approximately one-third of the master's degrees in computing.

However, when we look closer at the representation among women earning master's degrees in computing by race/ethnicity, as shown in Fig. 4.3, we see fewer degree attainment gaps between those historically minoritized groups in computing (see NCES Table 322.20). For instance, white and Asian women earn most degrees at the master's level across all fields and in computing. However, the degree attainment gap between Black and Hispanic/Latina women in computing observed at the bachelor's level is nearly eliminated among women earning master's degrees (see Fig. 4.3).

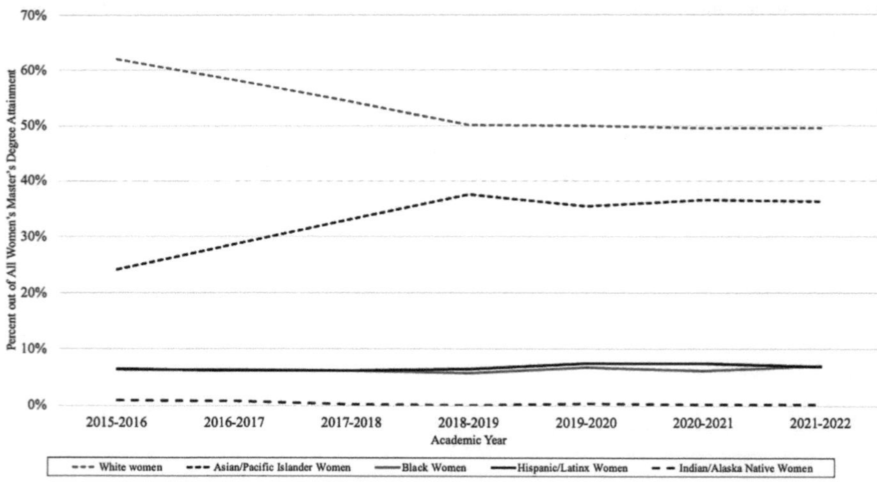

Fig. 4.3 Master's degrees awarded to women in computing by race/ethnicity: academic years 2015–2016 through 2021–2022. Note: This visual portrays trends from the last 7 years of available data on the percentage of master's degrees awarded to residential women by race/ethnicity. Data for this visual came from Table M7 in respective CRA Taulbee Reports from 2016–2022

PhD Level According to the 2021–2022 Taulbee Survey, there was an uptick in the number of students earning PhDs in computing (Zweben & Bizot, 2023). When investigating trends of PhD degree attainment over the last six decades, we can see that the proportion of degrees awarded to women has increased while the proportion of men earning PhDs has decreased (see Fig. 4.1). Despite some advancements toward gender equity, we continue to see gender gaps in the proportion of degrees awarded to men relative to women. For instance, in 1964, only 17% of all PhDs in computing were awarded to women, though for roughly the next 10 years, women earned fewer than 10% of all computing PhDs. Then, in the early 2000s, the percentage of women earning PhDs in computing rose to roughly a quarter—a rate that held fairly steady for approximately the last 20 years (see Fig. 4.1).

As is true for white women at the bachelor's and master's levels, white women earn the majority of PhDs across any field, followed by Asian/Pacific Islander women, Black women, and Hispanic women (31%, 7%, 6%, and 5%, respectively; see NCES Table 322.20). Among women who earn PhDs in computing, however, white and Asian women earn most of the doctoral degrees, following nonresident women, with the next largest groups being Black, Hispanic/Latina, and Indian/Alaska Native women (see Fig. 4.4).

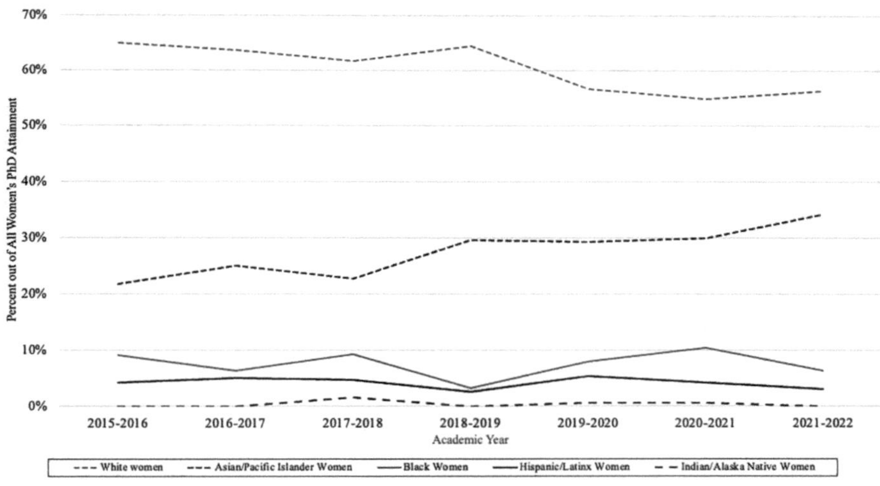

Fig. 4.4 PhDs awarded to women in computing by race/ethnicity: academic years 2015–2016 through 2021–2022. Note: This visual portrays trends from the last 7 years of available data on the percentage of master's degrees awarded to residential women by race/ethnicity. Data for this visual came from Table F8 in respective CRA Taulbee Reports from 2015 to 2021

4.5 Representation of Women in Academic Computing Positions

Like the trends shown above for women in academic computing at the undergraduate and graduate levels, we continue to see gender gaps among those in tenure-track academic computing positions. Using data from the Taulbee Survey reports, Fig. 4.5 demonstrates that the gender gap has only slightly narrowed between men and women across all tenure-track faculty positions over the last seven academic years (Zweben & Bizot, 2016, 2017, 2018, 2019, 2020, 2021, 2022, 2023). For instance, in 2015, only 15% of full professors were women, and as of 2022, only 17% of full professors are women. In 2015, only 23% of associate professors were women, and over the last 7 years, the representation of women has only increased by 1%. We have seen the most growth among assistant professors; in 2015, only 24% of assistant professors were women, and today, the percentage of women in this role has increased to 27%.

As shown in the sections above, all women—especially Women of Color—face disparities at each degree level in computing in terms of their representation in academic computing positions. Women with PhDs in engineering and computing fields are less likely than men to find employment after completing their degree (Kinoshita et al., 2020). Note that from 1977 until 1990, women and men with doctoral degrees in engineering and computing reported similar postgraduation employment levels. However, since the 1990s, the gender gap has grown such that in 2016, 22% of women earning doctorates in engineering/computing reported having no prospects after graduation, compared to 16% of men (Kinoshita et al., 2020). Women of Color reported higher levels of academic unemployment than did white women; in 2016, 19% of white women reported having no job prospects after earning their

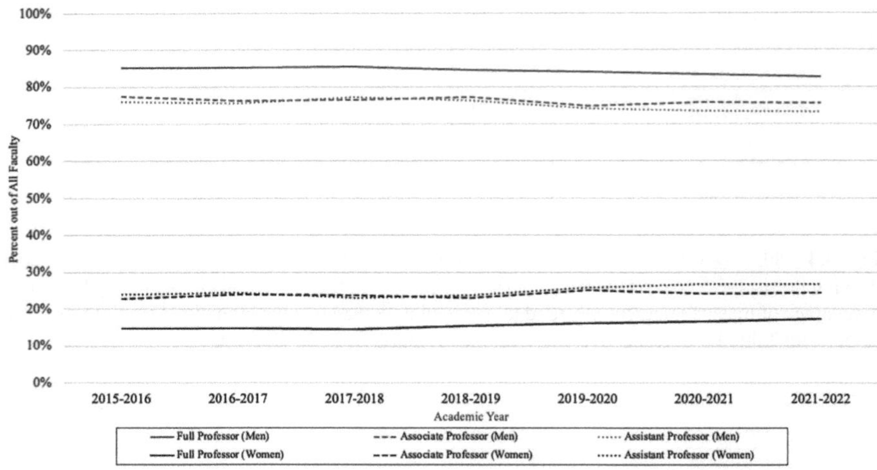

Fig. 4.5 Gender equity gaps of tenure-track faculty by academic rank, from 2015 to 2022. Note: Data for this visual came from Table F8 in respective CRA Taulbee Reports from 2015 to 2021

engineering/computing PhD, compared to 25% of Black women, 26% of Hispanic women, and 28% of Asian women (Kinoshita et al., 2020).

The challenges women face in the academic job market translate into lower representation in academic computing roles. According to the 2022 Taulbee Survey, women comprise 24% of all academic positions in computing, including tenure-track roles, teaching professors, academic researchers, postdoctoral scholars, and other academic roles (Dutt et al., 2016; Zweben & Bizot, 2023). Women's representation varies by role and level; women comprise over 30% of researchers and postdocs in computing (Zweben & Bizot, 2023). Among tenure-track faculty, women comprise approximately 17% of full professors versus about 27% of assistant professors in computing. Figure 4.5 illustrates the proportion of men's and women's representation in tenure-track computing faculty positions by rank. Women's representation across all ranks has been relatively stable from 2015 to 2022, with slight increases in women's representation at all levels.

4.6 Representation of Women of Color Among Computing Faculty

When the data reported by the Taulbee Survey are examined intersectionally, we see significant variation in women's representation by race/ethnicity and faculty rank, as shown in Table 4.2. Notably, white and Asian women make up the majority of women faculty at all ranks. There is somewhat greater racial/ethnic diversity among women at the assistant professor rank; however, this shift is mostly due to a

Table 4.2 Representation of women in computing faculty roles in 2021–2022, by race/ethnicity and rank

	Full professor		Associate professor		Assistant professor	
	# of total	% of total	# of total	% of total	# of total	% of total
Total faculty	1331	100	1168	100	1476	100
Total women	365	27	281	24	390	26
White women	238	18	148	13	172	12
Black women	7	1	16	1	18	1
Hispanic women	8	1	11	1	7	<1
Asian/Pacific Islander women	104	8	99	8	127	9
American Indian/Alaska Native women	4	<1	1	<1	3	<1
Women from two or more races	2	<1	2	<1	2	<1
Nonresident women	2	<1	4	<1	61	4

Note: The percentages for women by race/ethnicity are out of total faculty, not out of total women. Totals are calculated among those for whom gender and race/ethnicity are known. These data were adapted from CRA's 2022 Taulbee Report Table F8 (Zweben & Bizot, 2023)

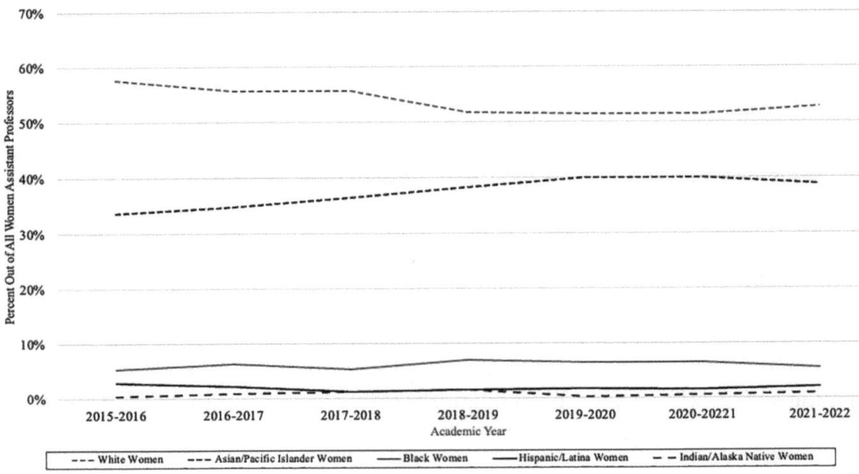

Fig. 4.6 Representation of women in computing assistant professor positions from 2015 to 2022, by race/ethnicity. Note: This visual portrays trends from the last 7 years of available data on the percentage of women in assistant professor positions by race/ethnicity. Data for this visual came from Table F8 in respective CRA Taulbee Reports from 2015 to 2021

relatively large proportion of nonresident faculty, for whom their racial/ethnic identities are unknown. Native American/American Indian/Native Hawaiian/Pacific Islander, African American/Black, and Hispanic/Latina women comprise 10% or less of all faculty and women faculty across the ranks.

The participation gaps for Women of Color in faculty roles are long-standing, with almost no change in the proportion of women assistant professors in computing who identify as Black, Hispanic/Latina, or American Indian/Alaska Native over the past 5 years (see Fig. 4.6). Further, the American Society for Engineering Education (ASEE) collects annual data on the demographics of faculty working in various disciplines that fall within engineering colleges in the United States, including computing. Analysis of ASEE longitudinal data revealed that over half of engineering schools did not employ a single Woman of Color in a faculty role until 2015 (Roy, 2022). Certainly, considerable effort and focus are required to address the significant inequities in women's participation, particularly among historically minoritized Women of Color, in computing faculty roles.

4.7 Navigating the Landscape of Women's Participation in Academic Computing

As described in the preceding sections, women have a long history of success in academic computing. Yet, we observe stubborn participation gaps for women at all levels of degree programs and academic computing roles, particularly when compared to women's overall representation in higher education. That said, some progress has been made in recent years, with modest gains in women's representation at the bachelor's, master's, and PhD levels. Most of these changes have been due to increases in white and/or Asian women's representation in computing. In contrast, the participation of women from other racially/ethnically minoritized groups has remained relatively stable. We have not yet seen the gains in women's participation in computing degree programs translate into increased representation among computing faculty ranks. Hence, it is unlikely that the participation gaps will resolve by themselves. Several organizations in computing seek to address these gaps and are actively working to support both individual women in their pursuit of careers in academic computing and address structures that have led to the inequalities in the first place. The following sections discuss the influences on women's low participation in academic computing, the types of interventions, and the organizations involved in these efforts.

4.8 Interventions and Organizations Addressing Women's Participation in Academic Computing Research Careers

4.8.1 Influences on Women's Low Presence in Academic Computing Research Careers

Participation of women in computing research careers often, but not always, proceeds along a common pathway: undergraduate research experiences; application and acceptance into graduate programs; support for graduate students and advisors in programs; placement into academic positions; and advancement in career hierarchies. We briefly summarize the nature of each stage of this journey.

Research Experiences for Undergraduates Undergraduates may not have had the opportunity to learn about research, research careers, and processes for pursuing research as a career. Common misconceptions include perceiving research as library-type research, where one compiles and summarizes others' work; belief that a master's degree is required for a PhD; and lack of awareness that research-oriented graduate school is similar to a paid apprenticeship (Barker, 2009; Cossa & Barker, 2021). When managed well, research experiences for undergraduates (REUs) are effective at cementing students' decisions to pursue graduate study and research careers, according to a 2010 meta-analysis (Sadler et al., 2010). Involvement in research increases women and racial/ethnic minority computing students' research skills, self-efficacy, scientific leadership, and scientific identity; however, the experience does not necessarily motivate students to pursue graduate study in the field (Rorrer et al., 2018). Many faculty reviewers for graduate degree programs look for research experience as evidence that their investment in a student is warranted. For these reasons, ensuring that undergraduate women learn about and participate in well-managed research experiences is an important first step in increasing the number of women in computing research careers.

Applications and Admissions Students who apply for research-based graduate programs are often ill-prepared to write effective personal statements or solicit strong recommendations from the right people, a factor especially true for students from underrepresented groups (Ramirez, 2011). Once applications are received, traditional admissions criteria such as scores on entry exams, grade point averages, and quality of the undergraduate institution can influence who is accepted. These metrics are shown to favor male, white, and Asian American students and to disadvantage older students, women, and other marginalized Students of Color (Bleske-Rechek & Browne, 2014; Miller & Stassun, 2014; Posselt, 2016). Faculty are often not trained to understand the biases implicit to recommendation letters, either in writing or in reading them. These are often written with a different "tone" for women than men, including subtle "doubt-raising" phrases that diminish women's strengths (Madera et al., 2018) or emphasizing non-research-related strengths over research ability. In addition, admissions committees can unknowingly admit applicants they perceive as similar to themselves (Posselt, 2016), a problem given the heavily male nature of computer science faculty. Increasing awareness of these issues among members of admissions committees and institutionalizing mechanisms to counter them are important for increasing women's participation in computing research.

Graduate Experience Once in a research-oriented graduate program, students must navigate several unfamiliar structures, including relationships with advisors, social ecosystem, and doctoral life cycle (NCWIT, 2021). The role of advisors is essential for women's success in doctoral studies. Advisors oversee students' progress through milestones, socialize them into local and disciplinary culture, and promote them in their networks (Posselt, 2018; Primé et al., 2015; Schlosser et al., 2011; van Rooij et al., 2019). Social ecosystems comprise the social climate experienced in coursework, labs, the department, and personal networks, which can contribute to feelings of isolation (McAlpine & Mitra, 2015; van Rooij et al., 2019). The gendered and racial/ethnic experiences of women, Students of Color, and first-generation students in graduate programs exacerbate these problems (Barker, 2016; Gardner & Holley, 2011; Joseph, 2012; Walkington, 2017). Finally, graduate students need different types of support at different stages of their doctoral journey and often do not understand what is needed of them at different times in their programs (Golde & Dore, 2001). Computer science departments that effectively manage all these concerns are rare.

Career Hierarchies To succeed in a research position in CS, faculty must compete for grants, present at conferences, and publish in conference proceedings and journals. As suggested in Part I of this chapter, the underrepresentation of women computer scientists in academic settings worsens at each step along the pathway, with women clustered in less influential positions. Several social factors help to explain the decline in advancement. Postdoctoral research positions are often a computer scientist's first position after earning a PhD. A large-scale study found that although women of all races and ethnicities are seriously considered and interviewed for science, technology, engineering, mathematical, and medical (STEMM) postdoc positions, white men are most likely to receive a job offer, followed by white women (Shauman & Huynh, 2023). Men and women computer scientists with postdoc experience are equally likely to be hired as assistant professors. Still, applicants with postdoc experience have more publications and, as a result, are more likely to be hired as assistant professors (Way et al., 2016). Women may also face hiring biases similar to graduate student admission, but the results are mixed (Ceci et al., 2023). Once hired, women computer scientists tend to produce fewer papers than men, which is worsened by and primarily explained by the "parenthood penalty" (Morgan et al., 2021). With respect to teaching, women computer scientists receive more negative ratings in teaching evaluations by students (Ceci et al., 2023; Gordon & Alam, 2021). Another tax on tenured women, and even more on Women of Color, is the invisible labor of service requests, including diversity efforts, mentoring, emotional labor, and maintaining a positive department climate (Gordon et al., 2022; Hanasono et al., 2019). Interventions on campuses, such as NSF ADVANCE programs, and special workshops, such as those held by the Computing Research Association, help women navigate these concerns.

4.9 Types of Interventions

Many efforts have been undertaken to reverse the underrepresentation described above. Generally speaking, two types of interventions are used to increase the participation of women in computing graduate programs and academic computing careers: individual-focused and systemic change or organizational approaches. Sometimes, individual-focused interventions use a deficit-based approach, which assumes that a member of a minority group enters an educational setting without the cultural knowledge or skills held by dominant group members (Yosso, 2005). Individual-focused approaches can also prepare and support women through the additional obstacles in their paths and do not assume that students lack skills. Deficit approaches frame men's experiences as a standard to which women are compared; when women fall short in knowledge, confidence, or cultural capital, programs furnish career advice, social support, ties to professional communities, and training so that women can cope within existing, male-dominated structures (Phillips & Beddoes, 2013; Vitores & Gil-Juárez, 2016). Some individual-focused approaches can be successful for the women participating in them but can have drawbacks. They can be expensive, such as sending students to conferences (e.g., Grace Hopper and Tapia conferences to celebrate diversity in computing). In addition, the interventions can tax the women involved by requiring them to divert energy, time, and attention away from research, advancing in their career as professionals, or participating in general departmental activities—a tax not required of their majority male peers. Publicity about special programs can also tacitly create or reinforce the perception that women are different, needy, less qualified, or getting an unfair advantage (Bierema, 2005; Convertino, 2020). Finally, individual-focused approaches do not involve the full range of stakeholders needed to positively impact academic women's experiences and, ultimately, success (Pifer & Baker, 2016). Most approaches to supporting academic women in computing fall into an individual-focused category, creating awareness among women of graduate careers, providing survival skills during graduate programs, and improving advancement from early entry into academic careers and onto senior administrative positions.

In contrast to individual-focused approaches, systemic change approaches are intended to change systems so all individuals within the system can thrive without additional support (Laursen & Austin, 2020; Reigeluth, 1992). Systemic change approaches do not assume that women in the system are deficient but instead cast social systems as historically serving the needs of men and being dominated by white masculine values and communication styles, thereby creating the perception or reality that women do not belong (Banchefsky & Park, 2018; Cabay et al., 2018; Weinberg et al., 2019). Systemic change interventionists evaluate the social structures of the organization and address culture, climate, learning environments, policies, and other structures that make entry to and advancement in doctoral programs and academic workplaces difficult for women. Because of the comprehensive nature of the changes required, systemic change approaches are very difficult and time-consuming. Prior researchers have suggested that in the existing state of gender inequity in organizations, we need multiple types of interventions, including those to equip women and those to change culture, policies, and practices (Kolb et al., 1998).

In the next section, we describe the interventions of six organizations to support academic career pathways.

4.10 Organizations and Their Interventions

Below, we describe six organizations' interventions and programs to advance women in computing research careers. A wide range of organizations are actively advancing women's and girls' participation in computing, at all levels of education and careers. Other organizations and programs address STEM women more generally, such as the National Science Foundation-funded ADVANCE program initiatives. One especially useful resource with systemic change strategies that has come out of NSF ADVANCE is the StratEGIC Toolkit (Strategies for Effecting Gender Equity and Institutional Change; StratEGIC Toolkit, 2018). We limit the discussion in this section to interventions and programs for women in computing research careers, selecting organizations with a primary focus on women in computing, resources and programs promoting women in computing research careers, and a national reach. We order the section starting with organizations with the greatest focus or most programs to advance women in computing research careers.

The organization that has been working in this space the longest and most comprehensively is the Computing Research Association, with research careers as its central mission. Other computing-focused organizations have varying efforts for academic (and industry/government) researchers. In the paragraphs below, we present only their efforts related to research careers and not their other initiatives.

4.10.1 Computing Research Association-Widening Participation

The Computing Research Association (CRA) is a nonprofit membership organization located in Washington, DC, that is centrally concerned with improving the research mission of computing. Part of that mission is to promote a diverse and equitable computing research community. In 1991, CRA established the Committee on the Status of Women in Computing (CRA-W), which has evolved into committees on Widening Participation and Education (CRA-WP and CRA-E; Aspray, 2016). Through these committees, CRA promotes research experiences for undergraduates, supports graduate students, and provides career advancement programming for women academics.

CRA-W began sponsoring two programs to provide research experiences for undergraduates (REUs) in 1994 (Barker, 2009). Today, they sponsor the National Science Foundation-funded Distributed Research Experiences for Undergraduates (DREU), a program that pays students competitively to work with faculty and graduate researchers at other institutions for 10 weeks in the summer. CRA provides a structure for DREU students, requiring the completion of several milestones, including progress and final reports. Faculty mentors are encouraged to participate in mentor training and are given tips for inclusive mentoring. Students who participate in the DREU program are consistently more likely to attend graduate programs than those who attend other REU programs or do not participate in REUs at all (Tamer, 2022). CRA also offers Scholarships for Women Studying Information Security (SWSIS). In 2024, CRA launched a new program, Undergraduate Research to PhD (UR2PhD), which provides a virtual undergraduate research methods course and mentoring advice for graduate students. UR2PhD also provides bridge workshops to support high-quality applications to graduate programs.

CRA has long been involved in studying and providing programming to improve women's graduate experiences. In 2001, CRA published a report describing experts' advice for recruiting and retaining graduate women in computing (Cuny & Aspray, 2001). The expert advice was later evaluated through the Graduate Recruitment and Retention program (Lord & Cohoon, 2007). Evidence was found to support several recommendations of the 2001 report concerned with systemic change: broaden admissions criteria, include diversity considerations in the admissions process, review departments' explicit and tacit messaging about programs, provide advice and support for mentors and advisors, improve official information, establish a departmental culture that values balancing work and life, and integrate graduate women into departmental culture and community. Since not all departments change in ways that support these admissions and retention mechanisms, CRA provides programs for retaining graduate students; the CRA's Grad Cohort for Women, launched in 2004, is a two-day workshop that provides women with graduate school survival skills, professional skills, networking experiences, and mentoring. Women or non-binary students who are also from underrepresented racial or ethnic groups or are living with disabilities can participate in the Grad Cohort for Inclusion, Diversity, Equity, Accessibility, and Leadership Skills (IDEALS), which was launched in 2018.

The first CRA-W academic faculty workshop for women was held in 1994, based on a workshop held at the 1993 Federated Computing Research Conference (Aspray, 2016). Today, Career Mentoring Workshops are offered to women annually or biannually. The workshops support senior doctoral students, early-career women (1–4 years after graduate school), and mid-career women (more than 5 years after graduate school) by providing professional advice, ranging across negotiating, building collaborations, teaching, and advancing into senior and leadership positions.

4.10.2 AnitaB.org

Anita Borg earned a PhD in computer science in 1981, focusing on operating systems. In 1987, she established an electronic mailing list called Systers to connect women in operating systems research (Aspray, 2016). Before establishing this list, women were isolated, often the only woman in a lab, and often unaware of other women until they attended conferences. Borg built an email system (Mecca) that allowed women to have a profile, similar to social networking sites today. Mecca both connected women computing researchers and allowed them to see themselves as members of a larger community. Later, Systers was opened to all technical women in any computing field and along any computing career path. Today, Systers claims more than 8500 members organized into 20 affinity groups worldwide. Except for two of these groups, the affinity groups are organized around national, ethnic, or racial identity and are no longer specific to the type of career.

AnitaB.org is best known for the Grace Hopper Celebration of Women in Computing (GHC), the world's largest gathering of women in technology. While GHC is lauded today for its massive career fair, one of GHC's focus areas is specifically for academics. Cognate organizations and individual speakers sponsor conference panels, talks, and workshops to support women in research careers. For example, CRA-WP held two "Career Conversations" at GHC 2023, one on how to apply to graduate school and the other on returning to graduate school after being in the workforce.

4.10.3 National Center for Women and IT (ncwit.org)

Founded in 2004, the National Center for Women and Information Technology, (NCWIT) is a national alliance of over 1500 organizations united to ensure that the perspectives and contributions of those who identify as girls or women are meaningfully represented at all levels of computing, including in research careers. With the exception of its Aspirations in Computing Programs, NCWIT primarily provides research-based resources for change leaders to make computing environments more equitable, but it also has some programs to support women in computing.

NCWIT promotes research experiences for undergraduates and provides guidance for managing them to members of its Higher Ed Alliance (formerly Academic Alliance). "REU-in-a-Box" is a comprehensive guide for initiating, managing, and completing a research experience that can be used by students and mentors (NCWIT, n.d.-b). Another resource summarizes the benefits of REUs for undergraduates, with two case studies to show faculty that they can support the advancement of women into research careers and that there are effective practices for including undergraduates in research lab activities and discussions.

NCWIT offers multiple award programs that support academic women in computing. The Harrold and Notkin Research and Graduate Mentoring Award is an annual award to recognize excellence in research, graduate mentoring, and diversity work. The NCWIT Aspirations in Computing Collegiate Award recognizes and draws attention to undergraduate and graduate women's technical accomplishments (NCWIT, n.d.-a). Finally, the NCWIT Higher Ed Alliance Seed Fund (supported by Microsoft Research) provides nonprofit US institutions with seed money to implement initiatives to recruit or retain women in computing.

NCWIT developed the Extension Services for Graduate Programs with seed funding from Google, Inc. This systemic change approach initiative, now named "Learning Circles," supports departments in self-assessing and then choosing research-based practices along all phases of graduate programs, including admissions, advising, doctoral life cycle, policies and accountability, social ecosystems, and ongoing evaluation. The resource collection includes several valuable resources, such as setting up a mentoring culture for graduate students and an overview of effective advising.

NCWIT has also collaborated with other organizations to support women in research careers. For example, Modern Figures, a collaboration with the Institute for African American Mentoring in Computing Sciences (iAAMCS), is a series of podcasts that draws attention to Black women's computing scholarship. Modern Figures has four seasons of conversational episodes that include technical talks, discussions of diversity, and survival skills for Black women in research careers (Modern Figures Podcast, 2017).

4.10.4 blackcomputeHER

blackcomputeHER.org was started in 2017 with the inaugural Black Women in Computing conference. The mission of the organization is to "address the lack of inclusion in the discipline by providing access to opportunity, education & training, and executive thought leadership" (blackcomputeher.org, n.d.).

The blackcomputeHER Fellows program supports Black women pursuing graduate studies in computing. This is a 12-month cohort-based program with social support and leadership development workshop-based professional learning opportunities. Each year's cohort is kicked off at the annual blackcomputeHER conference. Fellows in the program often go on in later years to serve as mentors to new cohorts of graduate students.

This organization also offers the CiteHER bibliography to make it easy for others to find great research by Black women in computing and technology and cite them in their work, supporting career advancement.

4.10.5 ACM-W

The Association for Computing Machinery (ACM) supports ACM-W, whose "core mission is to celebrate, inform, and support women in computing." ACM-W was founded in 1992–1993 and helps ACM members host regional celebrations of women in computing, connecting technical women in geographic areas.

Computer science research conference scholarships support women undergraduate and graduate students in computer science and related programs to attend computer science research conferences. This exposure to computing research can inspire a student to continue onto the next level of their academic or industry career. Student chapters support women in their undergraduate, graduate, and postgraduate goals by connecting them with peers and mentors. They receive guidance and resources from the global ACM community and can participate in Regional Celebrations of Women in Computing.

4.10.6 Alliance for Identity-Inclusive Computing Education

Funded in 2021 by a National Science Foundation Eddie Bernice Johnson INCLUDES grant, the Alliance for Identity-Inclusive Computing Education (AIICE) is focused on transforming K-16 education and supporting the identity-inclusive professional development of instructors and staff at the postsecondary level. AIICE often promotes policy-driven changes by providing programs to improve instructor and staff knowledge and skills to be inclusive educators.

The AIICE Teaching Assistant Professional Development Course was developed and piloted by AIICE partners Bard College and Mount Holyoke College. It aims to equip computer science teaching assistants (TAs) with the knowledge and tools to create an inclusive learning environment for students. The course supports independent learning by offering brief introductions to key topics such as mindset, emotional intelligence, effective feedback, identity, and belonging and then providing more resources for each topic to encourage deeper examination.

The 3C Fellows Program is a virtual professional development program where postsecondary (future) CS faculty and staff (in teams or individually) work in cohorts to "(1) study topics such as identity, intersectionality, and systemic disparities to understand historical trends that impact organizational cultures and technology development and (2) develop identity-inclusive computing courses, modules, and department activities" (Alliance for Identity-Inclusive Computing Education., n.d.). The first two cohorts' proposed deliverables vary in terms of improved onboarding of staff and students, new module development for advanced and introductory computing courses, and creating mentoring and community opportunities.

4.11 Evolving Strategies for Women's Inclusion in Academic Computing

Several organizations have addressed different aspects of women's underrepresentation in academic computing research since Anita Borg's creation of the Systers email list in the late 1980s. Many of these early efforts were designed to equip women or "fix the person" by creating a community for women, offering skills and professional development for individual women, and trying to lessen isolation. Current efforts also address the same issues through conferences, research experiences, professional development workshops, and leadership networking events.

More recent efforts have focused more on addressing structural or systemic issues. Multiple organizations discussed above have programs to support departmental change efforts. In addition, efforts such as podcasts and awards certainly have individual impacts on people's careers but also help change the dialogue and culture. These make visible the work of academic women in computing so that people who might have been relatively invisible before are no longer "hidden figures."

Addressing equity issues in academic computing requires efforts at both the individual and the organizational level. Many of the projects discussed in this section focus on undergraduate or graduate students and getting them interested and retained in academic computing research. Fewer projects support women faculty or academic researchers to excel and thrive in their computing work. This is a gap that should be addressed in future programming.

4.12 Conclusion: A Hopeful Horizon for Women in Academic Computing

In looking forward, we must be mindful that the computing and technology landscape is amid a major transformation driven by artificial intelligence (AI), quantum computing, and machine learning. As AI is drastically reshaping the industry as we know it, it becomes necessary to embed women's diversity in future innovations (Roopaei et al., 2021). The historical patterns we explored earlier in this chapter on women's involvement in academic computing highlight the pressing need to break down barriers and establish inclusive pathways for Women of Color. Since AI is inclusive only to those involved in its construction, diversifying academic computing will require collective efforts from all of us to foster women's active engagement and empowerment. Thus, the essential work of the key organizations committed to gender equity in computing will be necessary to advance women's involvement in academic research careers. The involvement of women can ultimately help reshape perspectives in transformational technology research, development, and leadership, thereby driving equity in technology and the greater social landscape (Roopaei et al., 2021). As we navigate this evolving frontier, our collective responsibility is to foster academic environments that embrace diversity, enabling all women to participate in shaping the future of computing and technology and fostering a world where innovation transcends gender boundaries.

References

Alliance for Identity-Inclusive Computing Education. (n.d.). Professional development constellation. Retrieved from https://identityincs.org/aiice-overview/professional-development-constellation

Ashcraft, C., McLain, B., & Eger, E. (2016). Women in tech: The facts. The National Center for Women and Information Technology (NCWIT). Retrieved from https://ncwit.org/resource/thefacts/

Aspray, W. (2016). Women and underrepresented minorities in computing. *Springer International Publishing Switzerland.* https://doi.org/10.5555/3279098

Banchefsky, S., & Park, B. (2018). Negative gender ideologies and gender-science stereotypes are more pervasive in male-dominated academic disciplines. *Social Science, 7*(2), Article 2. https://doi.org/10.3390/socsci7020027

Barker, L. (2009). Student and faculty perceptions of undergraduate research experiences in computing. *ACM Transactions on Computing Education., 9*(1), 1–28. https://doi.org/10.1145/1513593.1513598

Barker, M. J. (2016). The doctorate in black and white: Exploring the engagement of black doctoral students in cross race advising relationships with white faculty. *The Western Journal of Black Studies, 40*(2), 126–140.

Beyer, S. (2014). Why are women underrepresented in computer science? Gender differences in stereotypes, self-efficacy, values, and interests and predictors of future CS course-taking and grades. *Computer Science Education, 24*(2–3), 153–192.

Bierema, L. L. (2005). Women's networks: A career development intervention or impediment? *Human Resource Development International, 8*(2), 207–224. https://doi.org/10.1080/13678860500100517

blackcomputeHER. (n.d.). About. Retrieved from https://blackcomputeher.org/about-us/

Bleske-Rechek, A., & Browne, K. (2014). Trends in GRE scores and graduate enrollments by gender and ethnicity. *Intelligence, 46*, 25–34. https://doi.org/10.1016/j.intell.2014.05.005

Cabay, M., Bernstein, B. L., Rivers, M., & Fabert, N. (2018). Chilly climates, balancing acts, and shifting pathways: What happens to women in STEM doctoral programs. *Social Science, 7*(2), Article 2. https://doi.org/10.3390/socsci7020023

Ceci, S. J., Kahn, S., & Williams, W. M. (2023). Exploring gender bias in six key domains of academic science: An adversarial collaboration. *Psychological Science in the Public Interest, 24*(1), 15–73. https://doi.org/10.1177/15291006231163179

Convertino, C. (2020). Nuancing the discourse of underrepresentation: A feminist post-structural analysis of gender inequality in computer science education in the US. *Gender and Education, 32*(5), 594–607. https://doi.org/10.1080/09540253.2019.1632417

Cossa, J., & Barker, L. (2021). Motivation to pursue a Ph.D. in computing: Black students in HBCUs. *Journal of Interdisciplinary Studies in Education, 10*(1), Article 1. Retrieved from https://ojed.org/index.php/jise/article/view/2685

Cuny, J., & Aspray, W. (2001). *Recruitment and retention of women graduate students in computer science and engineering: Results of a workshop*. Computing Research Association. Retrieved from http://archive2.cra.org/uploads/documents/resources/workforce_history_reports/gradrr07.pdf

Dutt, K., Pfaff, D. L., Bernstein, A. F., Dillard, J. S., & Block, C. J. (2016). Gender differences in recommendation letters for postdoctoral fellowships in geoscience. *Nature Geoscience; London, 9*(11), 805–808. https://doi.org/10.1038/ngeo2819

Eagan, M. K., Stolzenberg, E. B., Zimmerman, H. B., Aragon, M. C., Whang Sayson, H., & Rios-Aguilar, C. (2017). *The American freshman: National norms fall 2016*. Higher Education Research Institute, UCLA.

Ensmenger, N. L. (2012). *The computer boys take over: Computers, programmers, and the politics of technical expertise*. MIT Press.

Fast facts: Women working in Academia. (2020). AAUW. Retrieved from https://www.aauw.org/resources/article/fast-facts-academia/

Gardner, S. K., & Holley, K. A. (2011). "Those invisible barriers are real": The progression of first-generation students through doctoral education. *Equity & Excellence in Education: University of Massachusetts School of Education Journal, 44*(1), 77–92.

Golde, C. M., & Dore, T. M. (2001). At cross purposes: What the experiences of today's doctoral students reveal about doctoral education. A report prepared for The Pew Charitable Trusts. Retrieved from http://www.phd-survey.org/

Gordon, H. R., Willink, K., & Hunter, K. (2022). Invisible labor and the associate professor: Identity and workload inequity. *Journal of Diversity in Higher Education, 17*(3), 285. https://doi.org/10.1037/dhe0000414

Gordon, N., & Alam, O. (2021). The role of race and gender in teaching evaluation of computer science professors: A large scale analysis on RateMyProfessor data. In *Proceedings of the 52nd ACM Technical Symposium on Computer Science Education* (pp. 980–986). https://doi.org/10.1145/3408877.3432369

Hanasono, L. K., Broido, E. M., Yacobucci, M. M., Root, K. V., Peña, S., & O'Neil, D. A. (2019). Secret service: Revealing gender biases in the visibility and value of faculty service. *Journal of Diversity in Higher Education, 12*(1), 85–98. https://doi.org/10.1037/dhe0000081

Joseph, J. (2012). From one culture to another: Years one and two of graduate school for African American women in the STEM fields. *International Journal of Doctoral Studies, 7*, 125–142.

Kinoshita, T. J., Knight, D. B., Borrego, M., & Wall Bortz, W. E. (2020). Illuminating systematic differences in no job offers for STEM doctoral recipients. *PLoS One, 15*(4), e0231567. https://doi.org/10.1371/journal.pone.0231567

Kolb, D., Fletcher, J. K., Merrill-Sands, D., & Ely, R. J. (1998). Making change: A framework for promoting gender equity in organizations. *CGO Insights, 1*, 1–4.

Lane, A., Mekonnen, R., Jang, C., Chen, P., & Lewis, C. M. (2021). Motivating literature and evaluation of the teaching practices game: Preparing teaching assistants to promote inclusivity. In *Proceedings of the 52nd ACM Technical Symposium on Computer Science Education* (pp. 816–822).

Laursen, S., & Austin, A. E. (2020). *Building Gender Equity in the Academy*. Johns Hopkins University Press. https://doi.org/10.1353/book.78724

Lord, H., & Cohoon, J. M. (2007). *Recruiting and retaining women graduate students in computer science and engineering*. In Report on research findings and workshop recommendations (p. 36). Computing Research Association. Retrieved from https://archive.cra.org/reports/gradrr07.pdf

Lunn, S., Zahedi, L., Ross, M., & Ohland, M. (2021). Exploration of intersectionality and computer science demographics: Understanding the historical context of shifts in participation. *ACM Transactions on Computing Education, 21*(2), 1–30.

Madera, J. M., Hebl, M. R., Dial, H., Martin, R., & Valian, V. (2018). Raising doubt in letters of recommendation for academia: Gender differences and their impact. *Journal of Business and Psychology, 34*, 287–303. https://doi.org/10.1007/s10869-018-9541-1

Main, J. B., Tan, L., Cox, M. F., McGee, E. O., & Katz, A. (2020). The correlation between undergraduate student diversity and the representation of women of color faculty in engineering. *Journal of Engineering Education, 109*(4), 843–864. https://doi.org/10.1002/jee.20361

McAlpine, L., & Mitra, M. (2015). Becoming a scientist: PhD workplaces and other sites of learning. *International Journal of Doctoral Studies, 10*, 111–128.

Miller, C. W., & Stassun, K. (2014). A test that fails. *Nature, 510*, 303–304.

Modern Figures Podcast. (July 22, 2017). Retrieved from https://modernfigurespodcast.com/about-us/

Morgan, A. C., Way, S. F., Hoefer, M. J. D., Larremore, D. B., Galesic, M., & Clauset, A. (2021). The unequal impact of parenthood in academia. *Science Advances, 7*(9), eabd1996. https://doi.org/10.1126/sciadv.abd1996

NCWIT. (June 11, 2021). Recruit and retain strategically for PhD programs. Retrieved from https://ncwit.org/resources/graduatesystemsmodel

NCWIT. (n.d.-a). NCWIT Aspirations in Computing Collegiate Award. Retrieved November 25, 2023, from https://www.aspirations.org/award-programs/aic-collegiate-award

NCWIT. (n.d.-b). REU-in-a-box: Expanding the pool of computing researchers. Retrieved November 25, 2023, from https://ncwit.org/resource/reubox/

Ong, M., Wright, C., Espinosa, L., & Orfield, G. (2011). Inside the double bind: A synthesis of empirical research on undergraduate and graduate women of color in science, technology, engineering, and mathematics. *Harvard Educational Review, 81*(2), 172–209.

Phillips, C. M. L., & Beddoes, K. (2013). Really changing the conversation: The deficit model and public understanding of engineering. 23.1025.1–23.1025.16. Retrieved from https://peer.asee.org/really-changing-the-conversation-the-deficit-model-and-public-understanding-of-engineering

Pifer, M. J., & Baker, V. L. (2016). Stage-based challenges and strategies for support in doctoral education: A practical guide for students, faculty members, and program administrators. *International Journal of Doctoral Studies, 11*, 015–034.

Posselt, J. (2018). Normalizing struggle: Dimensions of faculty support for doctoral students and implications for persistence and Well-being. *Journal of Higher Education, 89*(6), 988–1013. https://doi.org/10.1080/00221546.2018.1449080

Posselt, J. R. (2016). *Inside graduate admissions: Merit, diversity, and faculty gatekeeping*. Harvard University Press.

Primé, D. R., Bernstein, B. L., Wilkins, K. G., & Bekki, J. M. (2015). Measuring the advising alliance for female graduate students in science and engineering: An emerging structure. *Journal of Career Assessment, 23*(1), 64–78. https://doi.org/10.1177/1069072714523086

Ramirez, E. (2011). "No one taught me the steps": Latinos' experiences applying to graduate school. *Journal of Latinos and Education, 10*(3), 204–222. https://doi.org/10.1080/15348431.2011.581105

Reigeluth, C. M. (1992). The imperative for systemic change. *Educational Technology, 32*(11), 9–13. JSTOR.

Rodriguez, S. L., Lu, C., & Ramirez, D. (2020). Creating a conceptual framework for computing identity development for Latina undergraduate students. In *An asset-based approach to advancing latina students in STEM* (pp. 25–39). Routledge.

Roopaei, M., Horst, J., Klaas, E., Foster, G., Salmon-Stephens, T. J., & Grunow, J. (2021, May). Women in AI: barriers and solutions. In *2021 IEEE World AI IoT Congress (AIIoT)* (pp. 0497–0503). IEEE.

Rorrer, A. S., Allen, J., & Zuo, H. (2018). A national study of undergraduate research experiences in computing: implications for culturally relevant pedagogy. In *Proceedings of the 49th ACM Technical Symposium on Computer Science Education* (pp. 604–609). https://doi.org/10.1145/3159450.3159510

Roy, J. (June 1, 2022). *Faculty gender, race & ethnicity*. IRA. Retrieved from https://ira.asee.org/faculty-gender-race-ethnicity/

Sadler, T. D., Burgin, S., McKinney, L., & Ponjuan, L. (2010). Learning science through research apprenticeships: A critical review of the literature. *Journal of Research in Science Teaching, 47*(3), 235–256. https://doi.org/10.1002/tea.20326

Sax, L. J., Lehman, K. J., Jacobs, J. A., Kanny, M. A., Lim, G., Monje-Paulson, L., & Zimmerman, H. B. (2017). Anatomy of an enduring gender gap: The evolution of women's participation in computer science. *Journal of Higher Education, 88*(2), 258–293.

Schlosser, L. Z., Lyons, H. Z., Talleyrand, R. M., Kim, B. S. K., & Johnson, W. B. (2011). Advisor-advisee relationships in graduate training programs. *Journal of Career Development, 38*(1), 3–18. https://doi.org/10.1177/0894845309358887

Shauman, K. A., & Huynh, J. (2023). Gender, race-ethnicity and postdoctoral hiring in STEMM fields. *Social Science Research, 113*, 102854. https://doi.org/10.1016/j.ssresearch.2023.102854

Stolzenberg, E. B., Aragon, M. C., Romo, E., Couch, V., McLennan, D., Eagan, M. K., & Kang, N. (2020). *The American Freshman: National Norms Fall 2019*. Higher Education Research Institute, UCLA.

Stolzenberg, E. B., & Eagan, M. K. (2018). *Unpublished raw data on the American Freshman 2018 gender trends of those interested in majoring in computing*. Higher Education Research Institute.

Stolzenberg, E. B., Eagan, M. K., Aragon, M. C., Cesar-Davis, N. M., Jacobo, S., Couch, V., & Rios-Aguilar, C. (2019). *The American Freshman: National Norms Fall 2017*. Higher Education Research Institute, UCLA.

StratEGIC Toolkit. (March 27, 2018). Ethnography & evaluation research. Retrieved from https://www.colorado.edu/eer/research-areas/women-science/strategic-toolkit

Tamer, B. (2022). C/DREU students continue to pursue a graduate degree in computing at higher rates. *Computing Research News*. Retrieved from https://cra.org/crn/2022/01/c-dreu-students-continue-to-pursue-a-graduate-degree-in-computing-at-higher-rates/

van Rooij, E., Fokkens-Bruinsma, M., & Jansen, E. (2019). Factors that influence PhD candidates' success: The importance of PhD project characteristics. *Studies in Continuing Education, 43*, 48–67. https://doi.org/10.1080/0158037X.2019.1652158

Vitores, A., & Gil-Juárez, A. (2016). The trouble with 'women in computing': A critical examination of the deployment of research on the gender gap in computer science. *Journal of Gender Studies, 25*(6), 666–680. https://doi.org/10.1080/09589236.2015.1087309

Walkington, L. (2017). How far have we really come? Black women faculty and graduate students' experiences in higher education. *Humboldt Journal of Social Relations, 1*(39), 51–65.

Way, S. F., Larremore, D. B., & Clauset, A. (2016). Gender, productivity, and prestige in computer science faculty hiring networks. In *Proceedings of the 25th International Conference on World Wide Web* (pp. 1169–1179). https://doi.org/10.1145/2872427.2883073

Weinberg, F. J., Treviño, L. J., & Cleveland, A. O. (2019). Gendered communication and career outcomes: A construct validation and prediction of hierarchical advancement and non-hierarchical rewards. *Communication Research, 46*(4), 456–502. https://doi.org/10.1177/0093650215590605

Whitney, T. (2021). Women and People of Color still experience difficulty advancing in technology [Report].

Yosso, T. J. (2005). Whose culture has capital? A critical race theory discussion of community cultural wealth. *Race Ethnicity and Education, 8*(1), 69–91. https://doi.org/10.1080/1361332052000341006

Zweben, S., & Bizot, B. (2016). Continued booming undergraduate CS enrollment; Doctoral degree production dips slightly. *Computing Research Association*. Retrieved from https://cra.org/wp-content/uploads/2016/05/2015-Taulbee-Survey.pdf

Zweben, S., & Bizot, B. (2017). *Generation CS continues to produce record undergrad enrollment; Graduate degree production rises at both master's and doctoral levels. Computing Research Association*. Retrieved from https://cra.org/crn/wp-content/uploads/sites/7/2017/05/2016-Taulbee-Survey.pdf

Zweben, S., & Bizot, B. (2018). Another year of record undergrad enrollment; Doctoral degree production steady while master's production rises again. *Computing Research Association*. Retrieved from https://cra.org/wp-content/uploads/2018/05/2017-Taulbee-Survey-Report.pdf

Zweben, S., & Bizot, B. (2019). Undergrad enrollment continues upward; Doctoral degree production declines but doctoral enrollment rises. *Computing Research Association*. Retrieved from https://cra.org/wp-content/uploads/2019/05/2018_Taulbee_Survey.pdf

Zweben, S., & Bizot, B. (2020). Total undergrad cs enrollment rises again, but with fewer new majors; doctoral degree production recovers from last year's dip. *Computing Research Association*. Retrieved from https://cra.org/wp-content/uploads/2020/05/2019-Taulbee-Survey.pdf

Zweben, S., & Bizot, B. (2021). Bachelor's and doctoral degree production growth continues but new student enrollment shows declines. *Computing Research Association*. Retrieved from https://cra.org/wp-content/uploads/2021/05/2020-CRA-Taulbee-Survey.pdf

Zweben, S., & Bizot, B. (2022). CS enrollment grows at all degree levels, with increased gender diversity. *Computing Research Association*. Retrieved from https://cra.org/wp-content/uploads/2022/05/2021-Taulbee-Survey.pdf

Zweben, S., & Bizot, B. (2023). Record doctoral degree production: More increases in undergrad enrollment despite increased degree production. *Computing Research Association*. Retrieved from https://cra.org/crn/wp-content/uploads/sites/7/2023/05/2022-Taulbee-Survey-Final.pdf

Chapter 5
Women at the National Science Foundation

Linda G. Blevins and Emily G. Blevins

5.1 Introduction

The U.S. National Science Foundation (NSF) is an independent federal agency—with a fiscal year 2023 budget of about 10 billion dollars—that supports science and engineering research and education in all 50 states and US territories. The agency's investments make up about 25% of federal support to America's academic institutions for basic research. The NSF is organized into eight research directorates, including one for engineering (ENG) and one for computer and information science and engineering (CISE). The agency has about 2000 employees—including federal staff as well as rotators on leave from academic institutions. This chapter tells the story of academic women in engineering and computer science who have served the public as NSF ENG and CISE leaders, exploring their history, their present status, and the opportunities that will emerge for them in the future.

Much of the historical content in this chapter is drawn from two excellent books detailing the history of the formation of the NSF ENG directorate (Belanger, 1988) and the history of the NSF CISE directorate (Freeman et al., 2019). In addition, oral and/or written interviews were conducted specifically for this chapter by one of its co-authors, Linda Blevins. Some material comes from her personal recollections of her time served as Deputy Assistant Director for the ENG directorate from 2017 through 2021. Various other sources were consulted as cited in the text.

L. G. Blevins (✉)
U.S. National Science Foundation, Alexandria, VA, USA

E. G. Blevins
Alexandria, VA, USA

J. P. Carpenter (ed.), *Women in Academia*, Women in Engineering and Science,
https://doi.org/10.1007/978-3-031-97938-5_5

5.2 NSF History and Operations

On May 10, 1950, President Harry Truman signed the National Science Foundation Act, creating the first federal agency devoted to supporting fundamental research and education across all scientific disciplines. The act outlined the agency's core mission, which persists today: "To promote the progress of science; to advance the national health, prosperity, and welfare; and to secure the national defense; and for other purposes."

The act also stipulated that a presidentially appointed director would lead the agency and would share planning and decision-making responsibility with the National Science Board—a newly established advisory body made up of 24 representatives from the scientific community.

In terms of its initial organizational structure, the NSF Act stipulated that the agency have four operational divisions: a division for mathematical, physical, and engineering sciences (MPE); a division for medical sciences and a division for biological sciences, which quickly merged to become the division of biological and medical sciences (BMS); and a division for scientific personnel and education (SPE) to survey scientific manpower, promote science education, and oversee the awarding of college science fellowships.

Like science itself, however, in the nearly 75 years since its creation, NSF's programs and ambitions have never been static. Though its core mission remains largely unchanged, the composition of agency leadership as well as the agency's organizational structure have evolved and changed in response to advances in science and engineering, policy debates, and needs of US researchers, among many other factors.

Of particular relevance for this chapter, charting the agency's evolution over the past 75 years reveals an increasing programmatic focus on engineering and computer science research as well as the growing disciplinary representation among agency leadership. For example, as of 2024, there have been 15 presidentially appointed and senate-confirmed NSF directors (NSF, 2024). Of those 15 directors, 5 have been engineers or computer scientists:

- Dr. John B. Slaughter, engineer, 1980–1982
- Mr. Erich Bloch, engineer, 1984–1990
- Dr. Arden L. Bement, Jr., engineer, 2004–2010
- Dr. Subra Suresh, engineer, 2010–2013
- Dr. Sethuraman Panchanathan, computer scientist, 2020–present

Of NSF's 15 directors, only two have been women: Dr. Rita Colwell, microbiologist, 1998–2004, and Dr. France Córdova, astrophysicist, 2014–2020. As of this chapter's publication, NSF has yet to have at its helm an engineer or computer scientist who is also a woman. As subsequent sections of this chapter document, however, several women engineers and computer scientists have served in other leadership roles throughout the agency's history as well as today.

Engineering and computer science research have occupied increasingly prominent places within the agency's research focus over the past 75 years, which is reflected in the agency's changing structure. For example, the NSF Directorate for Engineering (ENG) was formed on March 8, 1981 (Belanger, 1988). Additionally, the NSF Directorate for Computer and Information Science and Engineering (CISE) was formed on May 1, 1986 (Freeman et al., 2019). CISE was initially comprised, in part, of some ENG programs and staff (along with programs and staff from other directorates). Most recently, in 2022, NSF formed a new directorate, Technology, Innovation, and Partnerships (TIP), initiated, in part, with a transfer of some ENG programs and staff. Many of the technologies emphasized by TIP rely on engineering and computer science expertise. As of 2022, NSF provides almost 80% of federal support for basic computer science research at institutions of higher education in the United States and about 30% of federal support for basic engineering research at those same institutions (NCSES, 2024b).

NSF's research directorates are each led by an Assistant Director (AD), who is also a member of the agency's senior executive leadership team. Technically, the people in these positions are titled the Assistant Directors of NSF. Referring to somebody as the AD for ENG, for example, is shorthand for AD of NSF for ENG. She is one of the eight assistant directors of the entire foundation but is the head of her specific directorate. Many ADs are rotators who come to NSF for up to 4 years on leave from a university using a government authority known as the Intergovernmental Personnel Act mobility program. Each AD has a (typically federal) executive Deputy AD or DAD who is a partner with the AD in strategic leadership and is also responsible for ensuring smooth directorate operations. In this chapter, fully appointed refers to an AD who is appointed by the NSF director to serve in the role without the Acting designation.

Out of 11 fully appointed ENG ADs, 2 have been women. The first woman AD for ENG joined NSF in 2017, 36 years after the directorate formed. Out of the 15 fully appointed CISE ADs, 3 have been women. The first woman AD for CISE joined NSF in 1998, 12 years after the directorate formed.

5.3 Past and Present Women at the Head of CISE and ENG

Five of the 26 fully appointed Assistant Directors (ADs) of the CISE and ENG directorates have been women. All have come to NSF on loan from academia. The first, Ruzena Bajcsy, began as AD for CISE in 1998, which is also the year the first woman director of NSF (Rita Colwell) started her position. The most recent, Susan Margulies, began as AD for ENG in 2021 and serves in that role as of the writing of this chapter in 2024. The collective story of all these women (including the two women directors of NSF) spans only the last 25 years of the agency's nearly 75-year history. All five of the women leaders of ENG and CISE are now members of the National Academy of Engineering (NAE), the most prestigious engineering honor in the nation. Remarkably, two of them are members of both NAE and the National Academy of Medicine (NAM), a dual honor that has only been bestowed on 90 individuals or 3% of NAE members (courtesy of NAE staff, 2024). They got their Ph.D.s in computer science or engineering over two decades (Bajcsy in 1972; Wing in 1983; Margulies in 1987; Martonosi in 1993; and Tilbury in 1996). They excelled in fields of study and professions where women were and continue to be underrepresented (Misa, 2010; see also Fig. 5.1 in this chapter). Their stories are told below.

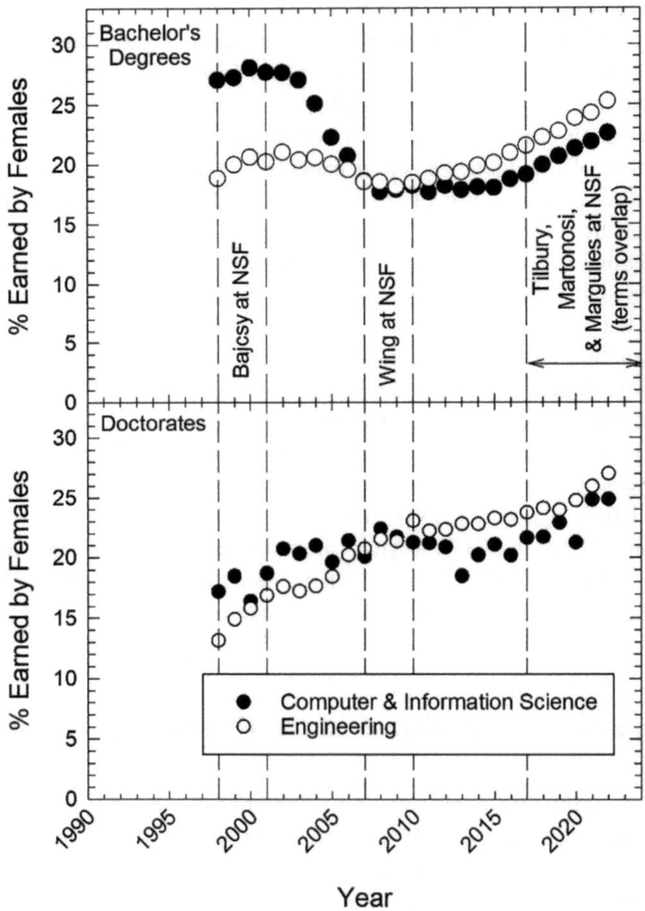

Fig. 5.1 Percent of bachelor's degrees and percent of doctorates awarded to women in engineering and in computer and information science from 1998 to 2022. The terms of the five women ADs are designated on the graphs by the vertical dashed lines (Department of Education, 2024; NCSES, 2024a)

5.3.1 Five Extraordinary Women Leaders

Ruzena K. Bajcsy, CISE AD, at NSF 1998–2001 (Fig. 5.2) Dr. Ruzena Bajcsy was born in Czechoslovakia in 1933 (Abbate, 2002). Her father was a civil engineer. Tragically, her parents were killed by the Nazis in 1944 when Ruzena was only 11 years old. Ruzena was then raised in orphanages and foster care. Inspired by her late father, Ruzena worked hard to be like him. She loved math and became the first woman to earn a Ph.D. in electrical engineering from Slovak Technical University in 1967. She travelled to California that year after being invited to explore the new field of computer science at Stanford, intending to stay only a year. While she was in America, the Soviets entered Czechoslovakia, causing her to make the wrenching

Fig. 5.2 Ruzena Bajcsy. Photo credit: Family Photo

decision to not return to her family, including two children, in Czechoslovakia. In 1972, Ruzena earned a second Ph.D.—this one in computer science—from Stanford University.

Dr. Bajcsy became a faculty member in 1972 at the University of Pennsylvania. When she joined the University of Pennsylvania, she was the only woman faculty member in her department. While at Penn, she served as Professor and Chair of Computer Science and Engineering. Her interests were initially in computer vision, but they evolved into computer perception and eventually robotics. She has always had an interest in psychology and the way humans and robots interact. This led her to become the Founder and Director of the General Robotics and Active Sensory Perception (GRASP) Laboratory. One project she is particularly proud of was creating a digital anatomy atlas of the brain. Dr. Bajcsy was elected to the National Academy of Medicine in 1995. She was elected to the National Academy of Engineering in 1997 for the development of active perception methods and for leadership in the community (NAE, 2024).

After nearly three decades at the University of Pennsylvania, Dr. Bajcsy joined NSF, where she was the first woman to serve as AD for NSF's CISE Directorate from 1998 to 2001. While at NSF, she was on leave from the University of Pennsylvania. She was the sixth AD for CISE (Freeman et al., 2019). She served under Dr. Rita Colwell, the first woman director of NSF. At NSF, she introduced the concept of cyberinfrastructure, led the initiation of the Information Technology Research (ITR) program spearheaded by the White House and infused with a new investment, and instituted CISE research awards of larger size with the goal of freeing investigators from frequent proposal writing to make them more productive at research (Freeman et al., 2019). Dr. Bajcsy cited the mentorship of Dr. Rita Colwell as being very important to her. She recalled Dr. Colwell saying, "Ruzena, I'm not a computer scientist, but I trust you." Bajcsy reflected, "... that was a tremendous comfort to me" (Bajcsy, 2023). She also praised the mentorship of Dr. Joseph Bordogna, who was NSF's Deputy Director at the time. She already knew Dr. Bordogna since he had been her Dean at the University of Pennsylvania. He had also previously served several years as the AD for the ENG directorate and undoubtedly shared his experiences freely with her.

Upon leaving NSF in 2001, she was recruited to become the Director of the Center for Information Technology Research in the Interest of Society (CITRIS) at the University of California, Berkeley, serving as Professor of Electrical Engineering and Computer Sciences and NEC chair holder. Dr. Bajcsy retired in 2021 but is still very active. She now happily serves as Professor Emerita at both the University of Pennsylvania and the University of California, Berkeley. To Ruzena, family is very important. Her children eventually came to the United States in the early 1990s. She is very proud that both her daughter and granddaughter are academic computer scientists. Ruzena's daughter, Dr. Klara Nahrstedt, is a professor of computer science at the University of Illinois Urbana Champaign and is currently the co-chair of CISE's Advisory Committee. Ruzena's granddaughter, Dr. Andrea Bajcsy, is an Assistant Professor of Computer Science at Carnegie Mellon University. Ruzena and Klara are the first mother-daughter pair to be members of the National Academy of Engineering (Seidlitz, 2022). Ružena was radiant on video when she talked about these two women, stating, "I am so proud. Look at me, how proud I am."

When interviewed for this book chapter, Dr. Bajcsy cited the importance of integrity and a moral compass in leadership in both academia and government. Her life story, as described above, gives her a strong belief system and a clear understanding of her priorities. When asked about NSF, she said, "I have the best memories of [my] time being there."

Jeannette M. Wing, CISE AD, at NSF 2007–2010 (Fig. 5.3) Dr. Jeannette Wing grew up in and around New York City. She loved mathematics and all kinds of logic and word puzzles as a little girl (Zgorski, 2022). When, in second grade, her teacher asked her what her father did, her mother taught her to memorize the sentence, "My father is a professor of electrical engineering at Columbia University." It was not until the eleventh grade that Jeannette asked her father what engineers do. Her dad told her that engineers use mathematics to solve real-world problems. From that moment forward, Jeannette wanted to study engineering. At MIT, as an electrical

Fig. 5.3 Jeannette Wing. Photo credit: Carnegie Mellon University

engineering major, she was required to take a computer science course. She fell in love with it, changed her major to computer science, and never looked back. She earned three degrees at MIT, culminating with her Ph.D. in computer science in 1983. As an MIT student, she interned at the Information Sciences Institute (ISI) at the University of Southern California (USC) and Xerox Palo Alto Research Centers.

Dr. Wing began her faculty career at USC, and, in 1985, she moved to Carnegie Mellon University (CMU), where she held various roles, including President's Professor of Computer Science, Department Head in Computer Science (twice), and Associate Dean for Academic Affairs of the School of Computer Science. During her sabbaticals, she was a visiting faculty member at MIT and later a visiting researcher at Microsoft Research. Her research contributions were in many fields but are most notably mentioned in formal methods, trustworthy computing, cybersecurity and privacy, programming languages, and software engineering. Dr. Wing, beginning with a 2006 essay in Communications of the ACM, popularized the term *computational thinking* to describe the thought processes that a computer scientist would use to solve a problem (Wing, 2006). She strongly believed that anyone and everyone could learn these techniques and could apply them in daily life (Microsoft, 2013).

Dr. Wing took leave from CMU to become the second woman to serve as AD for NSF's CISE directorate. Her NSF term lasted from 2007 to 2010. She was the eighth AD for CISE (Freeman et al., 2019). Dr. Wing wove her computational thinking concept into many of the programs she led at NSF, especially Cyber-enabled Discovery and Innovation (CDI). Dr. Wing is the proudest of (a) two cross-agency programs, CDI and what has now become Secure and Trustworthy Cyberspace (SaTC) and (b) the cross-directorate program Expeditions in Computing, designed to fund ambitious, cross-disciplinary, cross-institution research to define the future of computing. She promoted interdisciplinary programs within CISE, e.g., Smart and Connected Health (SCH), and programs run jointly with other directorates, e.g., with the ENG directorate, the Cyber-Physical Systems program, and, with the Social, Behavioral and Economic Sciences (SBE) directorate, the Social Computational Systems (SoCS) program and the Interface Between Computer Science and Economics and Social Science (ICES) program. She is also proud to have helped launch the Computing Community Consortium (CCC), an influential visioning organization for computing research. She also oversaw important collaborations with industry to make cloud software and services available to academic researchers, and she led CISE to partner with other directorates and agencies to create the National Robotics Initiative (NRI). Dr. Wing was a leader in NSF's administration of the historic American Recovery and Reinvestment Act of 2009 (Recovery Act) after the economic crisis of 2008. The Recovery Act gave NSF a one-time, 50% budget increase, which had to be spent quickly while under a high level of scrutiny (Freeman et al., 2019). When interviewed for this chapter (Wing, 2023), she said, "Being in a position like AD at CISE at NSF, you're ... helping make policy decisions all the time for the agency and you're balancing ... what's good for the science with other constraints. That kind of decision-making is something you're not taught or exposed to as an academic. My experience at NSF ... opened my eyes in many directions."

Upon leaving NSF, Dr. Wing returned to CMU. She took another leave from CMU to lead Microsoft Research's labs around the globe and was Corporate Vice President for Microsoft Research from 2013 to 2017. In 2017, she joined Columbia University as the inaugural Avanessians Director of the Data Science Institute. She is now Columbia's Executive Vice President for Research and Professor of Computer Science.

Dr. Wing was elected to the National Academy of Engineering in 2024 for formulation and advocacy of computational thinking and for contributions to formal methods and trustworthy computing (NAE, 2024). When asked to reflect on her time at NSF, she said it was "the highlight of my career."

Dawn M. Tilbury, ENG AD, at NSF 2017–2021 (Fig. 5.4) Dr. Dawn Tilbury grew up in a middle-class family in Minnesota. Her father was an electrical engineer at Honeywell and thought Dawn would be a good engineer, too. When she wanted to study French in college, her dad encouraged her to start in engineering while taking French electives to keep her options open. He worried that she would have to play catch-up if she wanted to change her major from French to engineering later.

Fig. 5.4 Dawn Tilbury. Photo credit: National Science Foundation

Later, Dawn expressed a preference for a physics major over an engineering one, and her father again nudged her toward keeping her options open by studying engineering, noting that engineers could get good jobs with undergraduate degrees while physicists would require graduate school. In the end, she took her father's advice and studied electrical engineering at the University of Minnesota, but she decided to go to graduate school after all. She moved to the University of California, Berkeley, earning a Ph.D. in electrical engineering and computer science in 1994 and facing a choice of whether to work in industry or academia. She again chose to keep her options open, assuming that it would be easier to move later from academia to industry than from industry to academia.

In 1995, Dr. Tilbury took a faculty position in mechanical engineering at the University of Michigan and, 30 years later, she is still a happy Michigan professor. Even though her doctorate was in electrical engineering, she was able to work in mechanical engineering because the field of control systems, her area of expertise, is cross-disciplinary (Šabanović, 2015). As a professor at the University of Michigan, she has spent time at several research institutes around the world (e.g., Italy, China, Sweden) and in industrial labs at IBM and DaimlerChrysler. Shortly after she joined Michigan, the NSF-funded Engineering Research Center for Reconfigurable Manufacturing Systems was launched, and she began applying her expertise in controls and robotics to problems in manufacturing. Her first academic leadership role was as a thrust leader in this NSF-funded center. She eventually became an Associate Department Chair and then the Associate Dean for Research in the College of Engineering. Notably, she was the inaugural chair of the Robotics Steering Committee.

Dr. Tilbury took leave from the University of Michigan to become the first woman to serve as NSF's AD for the ENG Directorate, serving from 2017 to 2021. She was the tenth AD for ENG. She was hired by Dr. France Córdova, the second woman director of NSF. When, in 2017, Dr. Tilbury hired Dr. Linda G. Blevins to serve as the Deputy Assistant Director for ENG, they became the first pair of women to serve as AD and DAD together for the ENG directorate.

Dr. Tilbury was at NSF during an eventful 4-year period. In an interview for this chapter, she described that, during her first year in the role, NSF moved its headquarters from Arlington, Virginia, to a brand-new building in Alexandria, Virginia. During her second year, NSF survived the longest-ever government shutdown of 5 weeks. In the third year, she says, "I famously told people: OK, after moving the building and the government shutdown, 2020 promises to be the perfect year. Nothing could possibly go that wrong" (Tilbury, 2024). A few weeks later, NSF staff rapidly transitioned to universal telework as the global pandemic unfolded. During her fourth year, there was a presidential transition and a great opportunity to educate the incoming administration about NSF. Dawn's last day at NSF was June 19, 2021, the first annual observance of the Juneteenth federal holiday.

From her time at NSF, Dr. Tilbury is most proud of a change she led to remove proposal submission deadlines in the directorate, encouraging investigators to submit whenever their ideas were ready rather than reactively when deadlines approached. The resulting reduction in proposal numbers reduced workload for the NSF staff, for the investigator community, and for the reviewers. Some of the new programs started under her leadership included Future Manufacturing, the Gen-IV Engineering Research Centers, and the Engineering Research Initiation (ERI) awards program. Dr. Tilbury is particularly proud of the ERI program because it focused on jump-starting the careers of early-career investigators at universities other than major research ones, thereby building capacity and diversifying engineering. She also led the launch of the Engineering Research Visioning Alliance (ERVA), a visioning organization for basic engineering research. Looking back, Dr. Tilbury valued her work with the other ADs to lead NSF's Ten Big Ideas (NSF, 2020). Although Dr. Tilbury did not create the Big Ideas, she carried them out, which is representative of the way initiatives that begin under one AD are often handed off and carried to fruition by the next AD (NSF ENG, 2021). Under her leadership, ENG took on stewardship for one of the Big Ideas, the Future of Work at the Human-Technology Frontier. She felt this was significant, commenting, "I think ... every single talk I gave talked about the Big Ideas and really communicating to the community that this is a new opportunity for us as academics to work together on problems of societal impact. ... that was something I really enjoyed participating in."

Dr. Tilbury returned to the University of Michigan when she left NSF. She went immediately into a new role of Associate Vice President for Convergence Research. She stayed in that role for about a year before she got a call from the Dean of Engineering, asking her to become inaugural Chair of the Robotics Department at Michigan. Dr. Tilbury had a history with this program: Back in 2012, she had led a committee that recommended that Michigan offer M.S. and Ph.D. degrees in Robotics, create a Robotics Research Institute, and build a new Robotics building. Michigan started the graduate programs in 2014 and formed the institute in 2017. A new building opened shortly before Dr. Tilbury's term at NSF ended. While she was at NSF, her colleagues at Michigan developed the new undergraduate program and proposed the creation of the department. She became inaugural Robotics Department Chair in 2022, 10 years after her committee had recommended a strategic focus on robotics for Michigan.

Dr. Tilbury was elected to the National Academy of Engineering in 2024 for advances in manufacturing network control and human-robot interaction and for engineering leadership (NAE, 2024).

Margaret R. Martonosi, CISE AD, at NSF 2020–2023 (Fig. 5.5) Dr. Margaret Martonosi's parents were both scientists and were the first in their families to go to college (Martonosi, 2024). Her father became a biochemistry professor, and her mother was a high school biology teacher. Margaret went to college thinking that she would major in psychology (CRA, 2013). She had enjoyed watching her younger sister learn as they grew up together and thought that psychology would help her understand what she had observed. However, as a freshman, she took and enjoyed a computer science course so much she decided to choose a major related to computers. Out of all the topics in computing, she most liked the idea of under-

Fig. 5.5 Margaret Martonosi. Photo credit: Princeton University

standing hardware, so she settled on electrical engineering. Though her parents were scientists, studying engineering was a way to forge her own, unique path in STEM. She earned a bachelor's degree from Cornell University and a master's degree from Stanford University, both in electrical engineering. She ultimately earned a Ph.D. in electrical engineering in 1993 from Stanford University.

Dr. Martonosi in 1994 took a faculty position at Princeton University, where she still works as the H.T. Adams '35 Professor of Computer Science. With Princeton as her home base, she spent a sabbatical at IBM as a visiting researcher. Dr. Martonosi is a computer architect. Her research studies have covered a wide span of topics such as the software/hardware interface, performance and power efficiency of devices, and mobile computing (CRA, 2013). Notably, she co-led with a campus biology collaborator the Princeton ZebraNet mobile sensor network project (Sampson, 2020). For ZebraNet, Dr. Martonosi's team designed and deployed in Kenya an energy-efficient sensing system featuring zebra tracking collars to study aspects of species endangerment and land use, expanding her interests from devices into systems. She served as Associate Dean for Princeton's School of Engineering and Applied Science and as Director of the Keller Center for Innovation in Engineering Education. She holds seven US patents.

Dr. Martonosi served from 2015 to 2016 as a Jefferson Science Fellow at the U.S. Department of State (Martonosi, 2017), where she applied her expertise to international computing and communications policy in fields as varied as the Internet of Things (IoT), technical standards and trade, financial technology, the Internet, security and privacy, artificial intelligence, and quantum computing. While at the State Department, she gained an appreciation of the need for more engineers to work in policy.

Four years after her stint at the State Department, Dr. Martonosi became the third woman to serve as AD for CISE. Her term lasted from 2020 until 2023. While at NSF, she was on leave from Princeton University. She was the 11th AD for CISE (Freeman et al., 2019). She began her term February 1, 2020, just a few weeks before the global pandemic was declared. Out of many accomplishments at NSF, she is most proud of her role in the COVID response, NSF's leadership in artificial intelligence (AI) investments and policy, and her ability to do virtual (during the pandemic) and in-person NSF outreach visits in all 50 states. During COVID, CISE quickly recognized the job market impacts and identified funding for the CI Fellows 2020 postdoc program to keep Ph.D.s on research pathways during the time of upheaval. It also doubled the amount of investment in undergraduate research to help students whose summer internships were being canceled; brought together CISE researchers working on COVID topics; and provided extra computing resources for the study of COVID. Both Drs. Martonosi and Tilbury were engaged in NSF's efforts under the historic Coronavirus Aid, Relief, and Economic Security Act (CARES Act) of 2020 and the American Rescue Plan Act of 2021 (ARP). Dr. Martonosi is proud that 25 new AI institutes were launched during her time at NSF, representing a total investment of half a billion dollars. She led her team to create structure and linkages among NSF's various investments in AI. Dr. Martonosi returned to Princeton University when she left NSF. She recently reflected on her time at NSF during a podcast interview (Gage, 2024):

> I came to the NSF because I was so appreciative of how it made so many aspects of my own professional life and career possible and beneficial for [my] students but for the country as well. I came to say thank you. When I got inside of NSF, I just realized how much more was going on. It's an amazing organization. I'm so proud to have gotten the chance to be a part of it. The number of lives that we're changing through offering opportunities for research is extraordinary, and the amount of real expertise and vision that gets brought to broadening that even further is something that amazes me.

Dr. Martonosi was elected to the National Academy of Engineering in 2021 for contributions to power-aware and power-efficient computer architectures and mobile systems (NAE, 2024).

Susan S. Margulies, ENG AD, at NSF 2021–Present (Fig. 5.6) Dr. Susan Margulies grew up in Minnesota. Her father was a clinician scientist, and her mother was a library scientist who created and ran a resource center at a local vocational institute. The existence of IBM and the Mayo Clinic in her hometown of Rochester, Minnesota, foreshadowed Susan's career performing research at the intersection of technology and medicine. She thought that she would be a premed student with a science or math major in college, but she learned about and fell in love with engineering at Princeton University. She earned a bachelor's degree in mechanical and aerospace engineering from Princeton University. She studied engineering with an eye on the machine that interested her the most, the human body. She went on to earn a Ph.D. in bioengineering and biomedical engineering in 1987 from the University of Pennsylvania, pursuing her interests to understand how rapid head movements can cause traumatic brain injuries. As a graduate student, she learned

Fig. 5.6 Susan Margulies. Photo credit: Emory University

about the many dimensions of faculty positions (research, teaching, mentoring, governance, service to scientific community) and was hooked: She chose a faculty career path and has not looked back.

While working as a postdoctoral trainee and research faculty member studying lung mechanics in health and disease in a clinical department at the Mayo Clinic, she enjoyed the transdisciplinary engineering, science, and medicine setting. After her postdoc, Dr. Margulies joined the faculty of the University of Pennsylvania in 1993 as the first woman faculty member in Bioengineering in the School of Engineering and Applied Sciences. She was the only woman faculty member in the department for 11 years. Throughout her career, she performed pioneering injury biomechanics research to determine the mechanisms of brain injuries in children and of lung injuries associated with mechanical ventilation with the aim of reducing injuries and helping with diagnosis and treatment (Hatch, 2006). In addition to her research impacts, she was actively engaged in creating campus-wide opportunities for faculty success. She was the Chair of the Faculty Senate at the University of Pennsylvania, and she was one of the six women faculty at the University of Pennsylvania who created the Penn Forum for Women Faculty, which still exists today.

In 2017, Dr. Margulies retired from Penn and became Professor and Chair of the Wallace H. Coulter Department of Biomedical Engineering (BME) at Georgia Institute of Technology and Emory University, leading a department that spans the two universities. She was the first faculty member tenured at both universities—in the College of Engineering at Georgia Tech and the School of Medicine at Emory. From her time in that role, she is proud to have launched many initiatives to solidify the inclusive culture of Coulter BME as One Department in Two Institutions. She built a one-department communications mindset, such that major awards to faculty on one campus were announced on both campuses. She built a unique philanthropy partnership, via establishment of a BME Common Fund, such that gifts to one institution could benefit either campus (e.g., an endowed chair gift could be made to one university while the chaired professor could be located at the other campus). She carried this focus on creative, inclusive partnerships to the NSF (University of Oregon, 2022).

Dr. Margulies became the second woman and first biomedical engineer to serve as AD for NSF's ENG directorate in August 2021 and, as of this writing, is still in the position. She is the 11th AD for ENG. While at NSF, she is on leave from her position as Georgia Research Alliance Eminent Scholar at Georgia Tech and Emory. She was alerted to the AD opportunity by Dr. Gilda Barabino, President of Olin College of Engineering, who was a member of the ENG directorate's Advisory Committee. Dr. Margulies was so inspired by the vision and enthusiasm of NSF's director, Dr. Panchanathan Sethuraman, and the compelling opportunity for engineering to impact the society, health, and economy of the nation that she decided to take the position. She loves teaching, mentoring, service, collaborating, partnering, continuous learning, and identifying new research opportunities. When asked what in her tenure (so far) makes her most proud, she chose to feature the ENG directorate's leadership in the agency-wide solicitations in Clean Energy Technology, modeled after ENG's stewardship of the Future of Work at the Human-Technology Frontier. For this program, ENG leads a cross-NSF working group that develops research topical areas and jointly decides how to invest new funds. Dr. Margulies is also proud to have started the Trailblazer Engineering Impact Award program, which is currently in its first year of competition. This award is focused on funding creative and innovative individuals who propose paradigm-shifting engineering research projects. While the high-risk, potentially high-reward program is just beginning, she hopes that it lives beyond her time as AD. While Dr. Margulies has always engaged in mentorship, she has learned more at NSF about being a role model and sponsor within NSF and for the communities served by NSF. Asked about her experience at NSF during an interview for this chapter, she stated, "... at the NSF, I am with like-minded individuals—where creating opportunities for others to achieve their potential is central to our shared mission" (Margulies, 2024).

Dr. Margulies was elected to the National Academy of Medicine in 2020. That same year, she was elected to the National Academy of Engineering for elaborating the traumatic injury thresholds of brain and lung in terms of structure-function mechanisms (NAE, 2024).

5.4 Academia and Government Symbiosis

5.4.1 *NSF as a Launch Pad into Academic Leadership*

The five pioneering ENG and CISE ADs described above were academics before joining NSF and either have or are planning to return to academia. However, there is another pathway into academia that involves NSF in a different way. The two women described in this section, Dr. Deborah Crawford and Dr. Grace Wang, began their careers in industry or in a combination of industry and a government lab—rather than in academia. They then came to NSF as program directors, managing the NSF proposal merit review process and making recommendations to their division leaders for which proposals to fund. They worked their way up to the NSF executive ranks to become Deputy Assistant Directors (DADs), one for CISE and the other for ENG. Both then were hired into academic executive roles, harnessing their experiences in industry and at NSF to lead academia into the future.

Deborah Crawford, CISE DAD, at NSF 1993–2010 (Fig. 5.7) Dr. Deborah Crawford grew up in a small mining town on the east side of Glasgow, Scotland (Hilton Hill, 2022). Her father was a civil engineer. She always loved math and science, even at the early age of 4 or 5, when her father gave her math problems to solve. She had an influential physics teacher in high school who believed in her and built up her self-confidence. She thought she would be a computer scientist, but she chose engineering because of the ability to solve bigger problems and collaborate in teams.

Deborah earned a bachelor's degree in electronic and electrical engineering from the University of Glasgow and went to work for a year as a researcher on solid-state lasers for General Electric Company at its Hirst Research Centre in London. While there, a professor at the University of Bradford in London who also had a visiting appointment at AT&T Bell Laboratories in Holmdel, New Jersey, asked her to be his

Fig. 5.7 Deborah Crawford. Photo credit: University of Tennessee

graduate student. Deborah got on a plane to New Jersey and immediately began her graduate research at Bell Labs. Three short years later, in 1989, she earned a Ph.D. in Information Systems Engineering from the University of Bradford in England. She was a postdoc at the University of California, Santa Barbara, and then she was a research engineer at Jet Propulsion Laboratory (JPL).

Dr. Crawford left JPL in 1993 for NSF. She served as a program director in several NSF organizations, including ENG, CISE, and Education and Human Resources (EHR, now called the Directorate for STEM Education or EDU), and as a senior advisor to the NSF Deputy Director, Dr. Joseph Bordogna. In 2002, she joined the NSF executive leadership ranks as DAD of the CISE directorate and served as Acting AD for CISE twice. She is most proud of being able to contribute in many ways at the agency, stating, "I trained as a device physicist, but I had the opportunity to help shape programs across many fields of STEM" (Crawford, 2024).

In 2010, Dr. Crawford left NSF to spend 4 years at Drexel University as Senior Vice Provost for Research; 2 years as President and Executive Director of the International Computer Science Institute; and 4 years as Vice President for Research, Innovation, and Economic Impact at George Mason University.

In 2020, Dr. Crawford became Vice Chancellor for Research, Innovation, and Economic Development at the University of Tennessee, and she remains in that position today. Throughout her career at NSF and in academia, she is most grateful for the relationships she has formed and kept. Asked about the similarities between NSF and academia, she states,

> NSF is a very well-run agency with a crystal-clear mission that drives everything the agency does. Because I had the opportunity to be involved in so many things at NSF, I really understood how the agency ran, which levers to pull to make things happen, to solve problems, to create results. I came to understand how the agency could operate at a high capacity because I understood its system of operations—the framework of policies and regulations, business process and procedures, stakeholder engagement and talent management and development—and importantly, their interdependencies—that create an effective organization. I brought that understanding to the universities I have been a part of, and I think it made a difference.

Grace Wang, ENG DAD, at NSF 2009–2016 (Fig. 5.8) Dr. Grace Wang started to be interested in science at the age of 9 (Wang, 2024a). Her mother, an accountant and executive, and her father, the founder and CEO of a pharmaceutical company, were very supportive. They gave Grace the space to pursue her passions, not pushing her toward any course of study. When Grace was in high school, her uncle published a book about polymer materials. The curious teenager (she did not know what a polymer was) read his book and, as a result, learned how engineering could enable real-world solutions to problems. She became very excited and eventually chose to study polymer materials in college. She earned her bachelor's and master's degrees in polymer materials. She then received a Ph.D. in materials science and engineering in 2001 from Northwestern University in Evanston, Illinois. Upon graduation, she joined IBM in California, doing research and development. Her areas of expertise at IBM/Hitachi Global Storage Technologies (HGST) were thin-film magnetic

Fig. 5.8 Grace Wang. Photo credit: Worcester Polytechnic Institute

recording media and carbon overcoat for data storage, and she holds seven US patents.

After 8 years at IBM/HGST, Dr. Wang took a job at NSF to be a program director in the Small Business Innovative Research/Small Business Technology Transfer Research (SBIR/STTR) program within the Industrial Innovation and Partnerships (IIP) Division of the ENG directorate. The SBIR/STTR program provides research and development funding to startup companies at the early stages of technology development. For context, the program started in 1977 at NSF and was so successful that it was expanded across the U.S. Government in 1982 (SBA, 2024).

Three years after joining NSF, Dr. Wang was promoted to the executive ranks as the first female IIP Division Director, providing strategic leadership over a variety of programs linking NSF's discovery research to industrial innovation for all of NSF. Within 2 years, she became the first woman DAD of the ENG directorate. After serving as Acting AD for a while, she joined academia. She served first as Vice Chancellor for Research and Economic Development at the State University of New York (SUNY) System and was soon promoted to Senior Vice Chancellor for Research and Economic Development. In parallel, she also served as Interim Provost for the SUNY System, and, for two and half years, she led SUNY Polytechnic Institute as Interim President. After her time at SUNY, she joined the Ohio State University to be the inaugural Executive Vice President for Research, Innovation, and Knowledge.

In 2023, Dr. Wang was appointed the 17th President of Worcester Polytechnic Institute (WPI). Thus, a career that started in industry turned into an academic one, made possible by her time at NSF. Dr. Wang is a thought leader on how universities can contribute to innovation, education, and economic growth (Wang, 2024c).

Dr. Wang is most proud of some of the programs she initiated or expanded at NSF. She observes that leading NSF and leading academia share some common characteristics—seeing the big picture and being a strategic thinker but also being practical enough to understand what resources and actions are needed to accomplish outcomes. The skills, she says, are transferable. She also points out the importance of consensus building in the cultures of both NSF and academia. Reflecting on her time at NSF, Dr. Wang said, "NSF is an amazingly motivational place, where one can think about the future of engineering research, education and innovation and figure out how to enable the best and the brightest in discovery, knowledge creation, and technology advancements" (Wang, 2024b).

5.4.2 *Women as Acting ADs*

During the rich histories of the ENG and CISE directorates, there have been several women Acting ADs who served in the time gaps between fully appointed ADs. These Acting ADs maintained continuity and momentum for the directorates. Most notably, Dr. Esin Gulari, Dean of Clemson's School of Engineering and Science, served as Acting AD for the ENG directorate from 2001 to 2003, 15 years before Dr. Tilbury was appointed as the first woman AD for the ENG directorate. Similarly, Drs. Crawford (2007) and Suzi Iacono (2014) served as Acting CISE AD, and Drs. Wang (2016) and Blevins (2021) served as Acting AD for ENG. This section features the story of one Acting AD and provides a fresh perspective on what it is like to serve in the position.

As of the writing of this chapter (Da Silva, 2024), Dr. Dilma Da Silva (Fig. 5.9) is the Acting CISE AD, and her career story adds new dimensions to those already told. Dr. Da Silva earned bachelor's and master's degrees in computer science in her native Brazil before earning a Ph.D. in computer science in 1996 at the Georgia Institute of Technology. She returned to Brazil to be a professor at the University of Sao Paulo for 5 years. After a successful 5 years in academia, in 2000, she joined IBM Research in New York, and she loved it. During her 12 years at IBM, she published and stayed active in service to the broader research community, which energized her personally but also kept her visible to academia. In 2012, she changed industry sectors, moving from working with large-scale supercomputers at IBM to working with small cell phones at Qualcomm in the Silicon Valley. She was recruited away from industry in 2014 to Texas A&M University to be Department Head of Computer Science and holder of the Ford Motor Company Design Professorship II. After serving 5 years as Chair (2014–2019), she served a year as Associate Dean for Faculty Success in the College of Engineering. While she could have continued moving up into academic leadership positions, she worked instead to establish herself as a more traditional faculty member in a US university—with a lab, grants, and teaching responsibilities—to open the door to greater career flexibility. In 2022, she went on a leave of absence from Texas A&M to join NSF as the Director of the Computing and Communications Fundamentals Division within CISE.

In late 2023, Dr. Martonosi returned to Princeton. When that happened, Dr. Da Silva was temporarily promoted from Division Director to Acting AD for CISE. Dr. Da Silva's experience represents a unique combination of time in industry, academia, and government. When asked about being an Acting AD, she pointed out that NSF benefits from having academics serve in leadership positions because they

Fig. 5.9 Dilma Da Silva. Photo credit: National Science Foundation

bring their unique perspectives as members of the research community that the agency is trying to serve. She values the ability to "shape the conversation" at the national level in talks with Congress, with the White House, and among those in the CISE research community (Da Silva, 2024). She also appreciates the ability to work alongside federal employees and points out that academics who serve as AD come with "fresh eyes" and that "... the fresh eyes, the different voices, the different experiences really enhance and make NSF stronger."

5.4.3 The Symbiosis

The eight women academics in engineering and computer science profiled in this chapter have served as leaders in academia and at NSF. However, they also worked in other sectors—in industrial research labs, in other government agencies and labs, and in clinical settings. Industrial research, when compared to most NSF-funded research, is typically more applied, seeks payback on shorter timescales, and is performed more for profitability than for the greater good and education of tomorrow's workforce. Industry research is also generally multidisciplinary, focused more on a product than a discipline. The women featured in this chapter have worked across disciplinary boundaries—in their research and in their time at NSF. While NSF is like academia in some ways, it is still a government agency with requirements to conform to the norms and expectations of stakeholders like the executive branch, Congress, and, ultimately, taxpayers. Despite the different emphases of these sectors, technology moves rapidly enough, and research is complex enough that time spent in all these sectors and working across disciplines makes one a better engineer or computer scientist. This section provides a few examples of the symbiotic relationships among sectors.

When the women featured in this chapter were asked how their prior experiences in academia had benefited NSF, several themes emerged. First, it was felt that academic leaders are well suited to lead at NSF because academia and NSF share a commitment to long-term, basic research and to preparing the workforce of tomorrow. NSF rotators who are active researchers bring with them their disciplinary expertise and knowledge of the latest research topics and trends. Faculty members with experience doing research, teaching, and performing community service can leverage this experience to craft NSF policies, procedures, and programs that make sense to those who must carry them out. The agency benefits from having leaders who know how to listen to the academic community. Further, academics bring experiences with partnering and collaboration into the agency. Finally, the women described here are all underrepresented in their fields. Having leaders with a diverse set of viewpoints and experiences brings breadth and depth to agency decision-making.

When asked how their experience at NSF had benefited academia, the responses were just as rich. Most commonly, these women cited a new appreciation for and skillset in big picture, horizon thinking and for viewing the needs of the country as a whole. They gained experience justifying and defending budget requests to the White House and to the U.S. Congress, learning in the process the importance of communicating about the societal benefits of research to stakeholders who are not necessarily scientists. They gained exposure to other federal agencies by participating in interagency councils and initiatives and learned how to build partnerships with industry. They learned how to collaborate across the wide range of research disciplines funded by NSF, how to work in teams, and how to make tough trade-offs in policy and budgeting. They gained appreciation for federal employees and how serious they are about spending taxpayer dollars wisely. They gained new appreciation for broadening participation in their fields. They learned how to be role models and sponsors. In moving from NSF to academia, they became ambassadors for the U.S. Government. Their time at NSF gave them credibility with other faculty to be able to prepare their colleagues for new government programs or policies. They sometimes took these skills into high-level jobs in university administration, where the scope of their influence broadened within their own universities but also through interactions with other universities across the nation via councils and boards. To a person, all of the women interviewed described the importance of the relationships they formed at NSF to their career success. These included relationships with other

ADs, with leadership mentors at all levels, and with the staff of their directorates and beyond.

In looking at their collective experiences across industry, academia, and government and their work across disciplines, a picture emerges of symbiosis. Leadership skills gained in one sector are transferable to the others. The ability to see the big picture, to think strategically, and to lead others with integrity, humility, kindness, and courage apply anywhere. Knowledge gained in each of these sectors strengthens experiences in the others, and these women are living proof that all sectors and disciplines are stronger when such movement occurs.

5.5 Opportunities for Future NSF Women Leaders

While all the leaders whose stories are told in this chapter have broken new ground, there are still leadership trails to be blazed at NSF for women in computer science and engineering. For example, as already pointed out, there has not yet been a woman engineer or computer scientist at the helm of the agency. An interesting observation is that the first woman CISE AD served under the first woman NSF director and the first woman ENG AD served under the second woman NSF director. The agency is certainly changing: While Dr. Bajcsy said that she was one of the two women ADs at the agency in the late 1990s, Dr. Margulies explained that women make up the majority of NSF ADs in 2024.

Further, there is an opportunity for future ADs for CISE and ENG to come from a wider variety of backgrounds. Interestingly, all five of the women who have been fully appointed as AD have had one or more parents in STEM. Impressively, all have attended and worked at elite, research-intensive universities. The door is wide open for future ADs who learned about STEM later in life or who come from other types of institutions. As Dr. Martonosi stated in a panel discussion (Zgorski, 2022), “My parents were both scientists; … I was immersed in that all along. ... And one of the things that we try to do at NSF is to make sure that, whether you [have STEM parents] or not, you’re seeing these opportunities and getting to build on them.” The future ADs of CISE and ENG will emerge from the pipeline being built right now and referenced by Dr. Martonosi.

Looking at the women whose stories are told in this chapter, it is clear that the ENG and CISE directorates are closely coupled. In academia, computer science departments are sometimes part of engineering colleges and schools. That the ENG and CISE directorates work closely together is no surprise. When CISE was formed in 1986, some ENG programs (Software Systems Design; Computer Systems Architecture; Vision, Robotic, and Knowledge-based Systems; and Communications and Signal Processing) and their staff moved to CISE to make up parts of the new directorate (Freeman et al., 2019). Notably, women electrical engineers have served at the helm of both ENG and CISE. However, there has not yet been a woman computer scientist at the head of ENG. This is another trail to blaze.

The close relationship between ENG and CISE now incorporates the new TIP directorate. When TIP was formed in 2022, both ENG and CISE were important contributors to its establishment. First, the AD for TIP had been a long-term executive in CISE, and the DAD for TIP had served as an executive in both ENG and CISE. Like when CISE was established, some ENG programs and their staff (including SBIR/STTR and I-Corps) were moved to make up a part of TIP. The fresh focus of TIP is creating new types of leadership opportunities for women academics in computer science and engineering in the future.

The TIP AD and DAD, Dr. Erwin Gianchandani and Ms. Gracie Narcho, respectively, were interviewed for this chapter. The new directorate was born at the nexus of Congressional, White House, and NSF priorities to bring focus to "... more use-inspired, solution-oriented, and translational science and engineering that is so critical to the Nation's innovation ecosystem," according to Gianchandani (Gianchandani & Narcho, 2024).

One unique aspect of TIP is the culture change it is pursuing, both in academia and inside NSF. Within academia, TIP wants to incentivize looking beyond publications as a primary measure of success, encouraging the use of other indicators such as patents, licenses, scaling of workforce development solutions, and so on. Inside of NSF, TIP is "trying to grow the types of expertise and the skill set of the workforce," according to Narcho. Looking at TIP's current talent (in 2024), it is made up of people from academia, industry, nonprofits, venture capital, startups, and technology transfer offices. According to Gianchandani, "We're trying to recruit a different cross section of folks to come work in TIP and trying to also give that talent exposure to different pathways going out of TIP." Another area of culture change is partnerships. According to Narcho, "We are trying to help the academic community appreciate the value of partnerships, partnerships with industry, partnerships with state and local government, and partnerships with angel and venture capital. Part of that is also to ... not have such a barrier for folks moving in and out of academia into some of these other areas." The women leaders described in this chapter are already demonstrating this, moving fluidly between industry, government, and academia and working across disciplinary boundaries. However, TIP is taking this to the next level, introducing other sectors and foci into the mix. Future leaders will participate in and be from these other sectors, too.

Academics in computer science and engineering may wonder how to build careers like those described here. There are myriad possibilities, as demonstrated in their varied stories. Notably, several of the women here got their first leadership opportunities in large, NSF-funded programs and centers, and several of them recommended this as a path forward. While this chapter has focused on the highest leadership positions of AD and DAD in the CISE and ENG directorates, there are other ways for academics to lead at NSF. Serving as an NSF Program Director rotator allows one to develop leadership skills within one's specialty research area. Serving as an NSF division-leading rotator allows one to gain experience across a broader set of subdisciplines. Serving as a rotator in some cross-cutting organizations like the Directorate for STEM Education (EDU), the Office of Integrative

Activities (OIA), the Office of International Science and Engineering (OISE), or TIP can offer boundary-spanning, cross-cutting perspectives. Further, serving on Advisory Committees allows one to advise NSF in a variety of programmatic areas. These are all rewarding opportunities to serve the computer science and engineering communities, NSF, and the nation and to build careers in leadership like the ones featured here.

To conclude, the women leaders of this chapter were asked about their dreams for the future of NSF. Some wished for NSF to continue to strengthen its curiosity-driven research, while others encouraged continued investment in technological innovation and stronger partnerships between academia and industry. A common theme was a hope for NSF to continue to support diversity and inclusion in engineering and computer science, emphasizing culture change over numbers and continuing to make NSF programs readily available to all. Finally, Dr. Bajcsy, who is 90 years old and has experienced the best and worst of the world, expressed a desire for less polarization and more collaboration in our society and in science. As Dr. Bajcsy states, "There are serious problems out there. And, for that, we need to stick together and use different skills and different parts of the brain to solve them. ... If we stick together, we will be able to solve them." The women in computer science and engineering whose journeys take them through NSF will lead the way in creating a better tomorrow.

Disclaimer The views expressed in this chapter do not necessarily reflect the views of the National Science Foundation or the United States.

References

Abbate, J. (9 July, 2002). Bajcsy, Ruzena, an oral history. Interview #575 for the IEEE History Center, The Institute of Electrical and Electronic Engineers, Inc. Retrieved from https://ethw.org/Oral-History:Ruzena_Bajcsy_(2002). Accessed 18 Feb 2024

Bajcsy, R. (2023). Interview. Conducted by Linda G. Blevins. December 6, 2023.

Belanger, D. O. (1988). *Enabling American Innovation: Engineering and the National Science Foundation*. Purdue University Press.

Computing Research Association's (CRA). (2013). Committee on Widening Participation in Computing Research. Summer/Fall 2013 Newsletter. *Interview with Margaret Martonosi*. Retrieved from https://cra.org/cra-wp/interview-with-margaret-martonosi/. Accessed 16 Feb 2024

Crawford, D. (2024). Personal Communication. Linda G. Blevins. January 11, 2024.

Da Silva, D. (2024). Interview. Conducted by Linda G. Blevins. January 29, 2024.

Department of Education. National Center for Education Statistics, Integrated Postsecondary Education Data System, Completions Survey, unrevised provisional release data. Downloaded 09 February 2024.

Freeman, P. A., Adrion, W. R., & Aspray, W. (2019). *Computing and the National Science Foundation, 1950–2016: Building a Foundation for Modern Computing*. Association for Computing Machinery.

Gage, S.L. (8 January, 2024). *Science Helps Us... Create the Future of Computing*. American Association for the Advancement of Science's Sci on the Fly. Retrieved from https://aaasstpf.libsyn.com/science-helps-us-create-the-future-of-computing. Accessed 16 Feb 2024
Gianchandani, E., & Narcho, G. (2024). Interview. Conducted by Linda G. Blevins. January 10, 2024.
Hatch, S. E. (2006). *Changing Our World: True Stories of Women Engineers* (p. 11). American Society of Civil Engineers.
Hilton Hill, H. (2022). Hallerin Hilton Hill Show: Interview with Deborah Crawford. Retrieved from https://research.utk.edu/oried/2022/10/04/hallerin-hilton-hill-radio-show-interview-with-deborah-crawford/. Accessed 16 Feb 2024
Margulies, S. (2024). Interview. Conducted by Linda G. Blevins. January 19, 2024.
Martonosi, M. (2017). Science, Policy, and Service: Some Thoughts on the Way Forward. Retrieved from https://www.cornell.edu/video/margaret-martonosi-science-policy-service. Accessed 16 Feb 2024
Martonosi, M. (2023). Personal Communication. Linda G. Blevins. November 28, 2023.
Martonosi, M. (2024). Personal Communication. Linda G. Blevins. February 20, 2024.
Microsoft. (13 October, 2013). *Luminaries: Jeannette Wing of Microsoft Research*. Microsoft Research Luminaries. Retrieved from https://learn.microsoft.com/en-us/shows/microsoft-research-luminaries/luminaries-jeannette-wing-of-microsoft-research. Accessed 17 Feb 2024
Misa, T. J. (2010). *Gender Codes*: Why Women are Leaving Computing. John Wiley & Sons, Inc.
National Academy of Engineering (NAE). (2024). Members Directory. Retrieved from https://www.nae.edu/20412/MemberDirectory. Accessed 29 Feb 2024
National Center for Science and Engineering Statistics. Survey of Earned Doctorates. Downloaded 09 February 2024a.
National Center for Science and Engineering Statistics. Survey of Federal Funds for Research & Development. Downloaded 29 July 2024b.
National Science Foundation. List of NSF Directors, 1950—present. Retrieved from https://www.nsf.gov/od/nsf-director-list/nsf-directors.jsp. Accessed 18 Feb 2024
National Science Foundation. (2020). NSF's 10 Big Ideas—Special Report. Retrieved from https://www.nsf.gov/news/special_reports/big_ideas/. Accessed 3 March 2024
National Science Foundation Engineering Directorate. (10 May, 2021). Four Decades of Creating the Future: A Celebration of the 40th Anniversary of the NSF Directorate for Engineering. Retrieved from https://players.brightcove.net/679256133001/NkgrDczuol_default/index.html?videoId=6261916875001. Accessed 18 Feb 2024
Šabanović, S. (22 January, 2015). Dawn Tilbury, an oral history. Interview # 795 for Indiana University and the IEEE History Center, The Institute of Electrical and Electronics Engineers, Inc. Retrieved from https://ethw.org/Oral-History:Dawn_Tilbury. Accessed 18 Feb 2024
Sampson, A. (2020). Ask me Anything with Margaret Martonosi. Retrieved from https://www.youtube.com/watch?v=ig5xmWoyo_I. Accessed 16 Feb 2024
Seidlitz, A. (18 May, 2022). *Family History: Nahrstedt, Bajcsy Share Stage as First Mother-Daughter Pair Elected to the National Academy of Engineering*. University of Illinois Urbana-Champaign News. Retrieved from https://siebelschool.illinois.edu/news/family-history-nahrstedt-bajcsy-share-stage-as-first-mother-daughter-pair-elected-to-the-national-academy-of-engineering. Accessed 16 Feb 2024
Small Business Administration (SBA). (2024). Birth & History of the SBIR Program. Retrieved from https://www.sbir.gov/tutorials/program-basics/tutorial-5. Accessed 1 March 2024
Tilbury, D. (2024). Interview. Conducted by Linda G. Blevins. January 26, 2024.
University of Oregon. (10 November, 2022). Knight Campus Distinguished Lecture. Susan Margulies: Transforming Health Through Engineering. Retrieved from https://www.youtube.com/watch?v=xmS6fngve00. Accessed 24 Feb 2024
Wang, G. (2023). Interview. Conducted by Linda G. Blevins. December 11, 2023.
Wang, G. (2024a). Personal Communication. Linda G. Blevins. February 11, 2024.

Wang, G. (2024b). Personal Communication. Linda G. Blevins. March 15, 2024

Wang, G. J. (2024c). Revisiting the Connection Between Innovation, Education, and Regional Economic Growth. *Issues Sci. Technol., 40*(2), 37–42.

Wing, J. (2023). Interview. Conducted by Linda G. Blevins. December 6, 2023.

Wing, J. M. (2006). Computational Thinking. *Commun. ACM, 49*(3), 33–35.

Zgorski, L.-J. (1 August, 2022). *Wisdom and insight from women leading in computer science and engineering*. National Science Foundation's Science Matters. Retrieved from https://new.nsf.gov/science-matters/wisdom-insight-women-leading-computer-science. Accessed 16 Feb 2024

Chapter 6
The National Science Foundation's ADVANCE Program

Jessie A. DeAro, Keri Ann Sather-Wagstaff, Dana M. Britton, and Chrystal A. S. Smith

6.1 Systemic Change for STEM Faculty Equity: Necessary for Diversity in STEM

The systemic focus of the National Science Foundation's (NSF) ADVANCE program makes it unique among federal STEM workforce and diversity programs. Other NSF programs provide funds to individual STEM faculty members for their research or career development. ADVANCE focuses on the institutions of higher education (IHEs) where they work. A key assumption of the ADVANCE theory of change is that more equitable academic workplaces facilitate the recruitment, retention, and success of a more diverse STEM faculty. For the NSF ADVANCE program, "gender equity" is understood as fairness and justice in the design and application of policies, processes, and practices and in the treatment of individuals.

At the time of the publication of this volume, the NSF ADVANCE program was archived by the NSF and a majority of ADVANCE grants were terminated in 2025. This chapter should be read as a history.

J. A. DeAro (✉)
U.S. National Science Foundation, Alexandria, VA, USA
e-mail: jdearo@nsf.gov

K. A. Sather-Wagstaff
Framingham State University, Framingham, MA, USA
e-mail: ksatherwagstaff@framingham.edu

D. M. Britton
Rutgers University, New Brunswick, NJ, USA
e-mail: dana.britton@rutgers.edu

C. A. S. Smith
University of Connecticut, Storrs, CT, USA
e-mail: chrystal.smith@uconn.edu

J. P. Carpenter (ed.), *Women in Academia*, Women in Engineering and Science,
https://doi.org/10.1007/978-3-031-97938-5_6

Though the primary focus of ADVANCE remained gender equity, the program had been broadened by social and behavioral science research that illuminated the ways that gender combines with other social identities to shape experience, access, inclusion, and ability to thrive in the STEM workforce. Since 2016, ADVANCE solicitations (which are NSF's mechanism for soliciting grant proposals) have called for proposals that work to address these multiple confounding axes of inequity, including gender identity, race, ethnicity, disability, and LGBTQIA+ status.

NSF's overall mission to promote the progress of science is closely linked to its efforts to foster diversity in the STEM academic workforce. Another guiding assumption of ADVANCE is that more diverse voices—particularly among STEM faculty—produce more innovative science. Through their own research and in their roles as expert reviewers, individual faculty members influence research funded by the NSF and other federal STEM mission agencies. Faculty members design and implement STEM curriculum and make decisions about whether their teaching will be informed by scientific research on inclusion and STEM learning. Faculty members are academic career role models who select and support students and set standards for STEM disciplinary culture and ethical research practices for students, postdoctoral scholars, and other faculty members. The messages they communicate about the importance of an inclusive STEM culture (or not) influence mentees' career and educational decisions. Many STEM faculty members serve in leadership positions in higher education, STEM professional societies, editorial boards, accreditation working groups and in state and US government positions. In each of these domains, a lack of diversity inhibits the potential for discovery and innovation.

For decades, the prevailing explanation for a lack of diversity in the STEM workforce blamed individuals, arguing that women and minoritized students lacked interest or the aptitude to pursue and excel in the sciences. NSF ADVANCE strategies, on the other hand, draw on decades of social and behavioral science research demonstrating that lack of diversity is in large part the result of systemic factors embedded in institutions. These systemic factors include toxic cultures, inequitable and unclear practices, and cultural assumptions that shape our understanding of gender and the "ideal" scientist (Valian, 1998; Stewart & Valian, 2018). One recent study (Spoon et al., 2023) of nearly 250,000 US academics underlines this point. Spoon et al., 2023 found that women were more likely to leave academia than men in both STEM (including medicine) and non-STEM fields. Women described being "pushed" out by toxic workplace climates characterized by "dysfunctional departmental culture or leadership, harassment, or [a] feeling like they do not belong or fit" (p. 6). Men, on the other hand, described being "pulled" from academia by better opportunities. Even among men, however, work environments ranked second as a reason for leaving (Spoon et al., 2023). This suggests that addressing toxic workplace culture could benefit everyone.

The NSF ADVANCE program would have been 25 years old in 2026. The program was inaugurated in 2001, and grew out of prior NSF efforts to increase the participation of women in STEM academic research (Rosser, 2001). Between 1982 and 2000, the NSF had a series of six different programs that supported the research activities of women in STEM (Table 6.1). These programs were initiated in response to the 1980 Science and Engineering Equal Opportunities Act (42 U.S.C. §1885), which authorized the federal government, including NSF, to "… encourage men and women, equally, of all ethnic, racial, and economic backgrounds, including persons with disabilities, to acquire skills in science, engineering, and mathematics, to have equal opportunity in education, training, and employment in S&E fields, and thereby to promote S&E literacy and the full use of the human resources of the Nation in S&E." While NSF programs to support individuals were often very important to those individuals, many indicated that the academic workplace culture and climate were unwelcoming and the path to career success was not transparent or consistent (Rosser, 2001). These observations, as well as the conclusions in the 1999 Massachusetts Institute of Technology's "A Study on the Status of Women Faculty in Science at MIT," informed the development of the NSF ADVANCE program. The *Institutional Transformation* track focused on the implementation of systemic change in institutions of higher education to address workplace equity issues for STEM faculty. This is the only track that was included for the entire 25-year run of the program. The importance and success of this track resulted in the development of a *Catalyst* track to support institutional data collection and analysis work, especially important for IHEs without the infrastructure to collect adequate data on STEM faculty to identify and understand areas of potential inequity. The *Adaptation* track supported smaller scale systemic change projects that adapted strategies developed and tested by past ADVANCE awardees. Other Excellent accounts of the origins and history of the NSF ADVANCE program can be found in several articles and papers (Rosser, 2004, 2012; Stewart et al., 2007; DeAro et al., 2019). In this chapter, we aim to share examples of learning *from* the ADVANCE grantees, the

Table 6.1 History of the NSF gender-focused programs. Although the GSE Program ended in 2014, investments in research on gender and STEM education and workforce continued to be supported within the NSF's Directorate of STEM Education Core Research program (ECR) as one of many potential topics of STEM education, broadening participation, and workforce research, until 2025

History of NSF gender programs	Dates
Visiting Professorships for Women in Science and Engineering program (VPW)	1982–1997
Research Opportunities for Women program (ROW)	1985–1997
Research Planning Grants and Career Advancement Awards for Women Scientists and Engineers programs (RPG & CAA)	1990–1997
Faculty Awards for Women Scientists and Engineers program (FAW)	1991–1997
Program for Women and Girls (PWG) (later called the Program for Gender Equity (PGE), Gender Diversity in Science and Engineering (GDSE), and the Research on Gender in Science and Engineering (GSE) program)	1993–2014
Professional Opportunities for Women in Research and Education (POWRE) (replaced VPW, ROW, RPG, CAA, and FAW)	1997–2000
ADVANCE: Increasing the Participation and Advancement of Women in STEM Academic Careers	2001–2025

emerging learning *about* the ADVANCE program, as well as learning *by* the ADVANCE program that helped to refine program design and focus.

6.2 Learning from, About, and by the NSF ADVANCE Program

In the more than 20 years of its existence, the ADVANCE program has generated an enormous amount of knowledge about gender equity in STEM higher education. At its inception, ADVANCE took a radical step in framing gender inequity as an institutional problem in need of institutional solutions. For many early ADVANCE teams, it was the process of applying for an ADVANCE Institutional Transformation (IT) award that first made gender inequities evident to administrators and faculty members on their campuses. For other campuses, ADVANCE funds were critical to the creation of offices to collect and analyze institutional data to make visible the existence of gender-based inequities as well as monitor equity over time.

The lessons learned can be divided into three categories. The first category consists of lessons *from* ADVANCE. This includes the toolkit of strategies developed and employed by ADVANCE institutions and the accumulated wisdom about which strategies were effective, which were not, and why. The second category is composed of lessons learned *about* ADVANCE. This includes research on the program's impact and reach. The third category contains lessons learned *by* the ADVANCE program. These are lessons that resulted in administrative changes to the ADVANCE solicitation, funding practices and priorities, and public-facing statements by those who administered the program (Laursen & De Welde, 2019; DeAro et al., 2019). Many of these changes reflect a broader, richer vision of systemic gender inequity than that framed by the original solicitation. Others have come from a desire to extend the reach of limited funding to more and different types of higher education institutions (Laursen & De Welde, 2019; DeAro et al., 2019).

6.3 Learning from the NSF ADVANCE Program

The first book to collect lessons learned from the original cohorts of ADVANCE awardees, *Transforming Science and Engineering*, edited by Abigail Stewart, Janet Malley, and Danielle Lavaque-Manty, appeared in 2007. Many of the programs that would come to be most associated with ADVANCE appear in these chapters—for example, the University of Michigan's Strategies and Tactics for Recruiting to Improve Diversity and Excellence (STRIDE) program for faculty recruitment (#0123571, https://advance.umich.edu/stride/). In 2012, Diana Bilimoria and Xiangfen Liang published the first evaluative study of ADVANCE programs, *Gender Equity in Science and Engineering: Advancing Change in Higher Education*, in which they identified five factors conducive to the success and institutionalization of institutional change initiatives: (1) senior administrative support and involvement, (2) a transformation champion, (3) collaborative leadership [of the ADVANCE project], (4) widespread and synergistic participation across units, and (5) visible actions (Bilimoria & Liang, 2012, p. 63).

The most recent book on ADVANCE, *Building Gender Equity in the Academy*, by Sandra Laursen and Ann Austin, was published in 2020. This work takes advantage of the perspective of 17 years of institutional change efforts and provides a synthetic overview of the types of strategies undertaken by ADVANCE grantees. In their review, Laursen and Austin (2020) emphasize that ADVANCE never intended to offer "one size fits all" solutions. The ADVANCE solicitation recognized the various challenges faced by institutions and encouraged teams to identify the problems their institution faced and to identify the most effective solutions for their context.

Drawing on 12 case studies of ADVANCE Institutional Transformation projects from cohorts 1 and 2, Laursen and Austin categorize ADVANCE systemic change strategies into four groups (Table 6.2). The two groups under "organizational strategies" are well aligned with the systemic change orientation of ADVANCE. Certainly, this is the case for organizational strategies focusing on policies, practices, and climate. So too are the initiatives in the "support the whole person" category, which also focus on changes in policy and practices related to dual-career and work-family issues.

At first glance, the "foster individual success" category does not fit the ADVANCE emphasis on systemic change. By definition, mentoring focuses on individuals, providing information and support to help them navigate their careers. Even so, ADVANCE funded these strategies—so long as they were only one component of an institution's strategies for change. Laursen and Austin (2020) argue that programs of this type remain necessary on the path to institutional change. They note that systemic and cultural changes may take decades to alter institutional practices and STEM cultures that have developed over centuries. However, the ADVANCE program solicitations explicitly required projects to include systemic changes to

Table 6.2 Classification of ADVANCE strategies. (Adapted from Laursen & Austin, 2020)

Organizational strategies	Institutional support for individuals
Interrupt biased processes: • Inclusive recruitment and hiring • Equitable processes of tenure and promotion • Stronger accountability structures	*Support the whole person:* • Support for dual-career couples • Flexible work arrangements • Practical family-friendly accommodations
Reboot the workplace: • Development of institutional leaders • Approaches to improve departmental climate • Enhanced visibility for women and gender issues	*Foster individual success:* • Faculty professional development • Awards to individual faculty • Mentoring and networking activities

address the underlying issues that cause the need for programs like networking and mentoring.

An example of a networking and mentoring effort that was implemented as part of an Institutional Transformation award to Kansas State University (KSU) in 2003 is the "ADVANCE Distinguished Lectureship" series (#0244984). The KSU program provided funds to junior women STEM faculty to invite a senior faculty member in their research area to campus to give an address and provide a few days of one-on-one mentorship. The idea was to build the networks of junior women to increase their access to opportunities for collaboration and to build their professional networks in ways that would contribute to their success in the tenure process. Preliminary data suggested that the program was successful in creating research opportunities and improving promotion and retention of women in STEM (Dyer & Montelone, 2007). Perhaps a greater testament to its success is that KSU institutionalized the networking program after the ADVANCE grant ended. Twenty years after the program started with ADVANCE support, KSU continues to fund faculty. The program has expanded to include associate-level faculty and is no longer limited to women (https://www.k-state.edu/kawse/advance/ADLS.html, accessed January 2024). This was only one among many initiatives KSU carried out—the full proposal included institutional strategies such as policy reviews and initiatives to increase the recruitment and retention of women faculty.

Another example comes from the Jackson State University (JSU) IT award (#1008549), which implemented summer writing retreats (JSU, n.d.-a). While these were very successful in producing grant proposals (25% success rate for funding) and peer-reviewed papers (60% publication rate), they also developed an institution-wide community for participants who may have been the only women in a STEM department. Loretta Moore, the principal investigator of the 2010 ADVANCE IT project at JSU, was also the first permanent woman chair in the College of Science, Engineering, and Technology at JSU in 1998. She later went on to become Vice President for Research and Federal Relations at JSU and to serve at NSF as the Section Head of the Established Program to Stimulate Competitive Research (EPSCoR) office. In fact, according to Moore, the lack of women in leadership roles was the motivation behind the writing retreats. In turn, these retreats helped to promote women in STEM who had stalled at the associate faculty rank (personal communication, January 30, 2024). This individual component was complemented with significant policy and structural changes, including a Diversity and Equal Employment Opportunity Office and the hiring of the first chief diversity officer in 2012. JSU also conducted gender equity and implicit bias training; and created JSU's first strategic plan for diversity (JSU, n.d.-b).

Teams funded by ADVANCE to implement mentoring programs have developed and tested new approaches. For example, ADVANCE grantees have developed group-based mentoring approaches that involve department chairs and faculty within and outside the department in "launch committees" for new faculty (Stewart et al., 2021; Michigan Tech, n.d.). Drawing on social science research, such programs usually include training for mentors and mentees and financial or workload incentives. Some of the important design factors of successful faculty mentoring programs in the ADVANCE portfolio are summarized in Table 6.3.

Over the years, many ADVANCE awardees expanded mentoring programs initially directed to women faculty to make them more broadly available, e.g., for all associate-rank faculty. Programs like this have shifted focus to include a wider spectrum of inequities that intersect with gender, including race and ethnicity, disability, and sexual orientation. When this emphasis on inequities was infused in mentor training, ADVANCE teams found that mentoring can become a vehicle for institutional change. As Thomas et al. observed of work done at Ohio State University (IT award #0811123): "[Our] workshop for [mentoring] circle participants and administrators led subsequently to college-wide teams that addressed problems identified in the circles. We conclude that peer mentoring as a means to facilitate institutional change has great potential" (Thomas et al., 2015, p. 143).

Although ADVANCE funded components to "foster individual success," the bottom-line orientation of ADVANCE did not change. The solicitations and funding practices continued to emphasize structural solutions as the most important priorities. The ADVANCE philosophy is that no amount of mentoring, networking, or direct support of individual faculty can make up for the inequities resulting from the lack of clear, transparent, and consistent policies and procedures related to faculty searches and hiring, teaching and service workload distribution, work-life programs and procedures, and tenure and promotion.

ADVANCE projects have been the subject of multiple books and hundreds of publications, many focused on the first two cohorts of Institutional Transformation grantees (19 IHEs in total). Rather than summarizing these previous publications, in this section we provide some examples of high-impact projects that exemplify ADVANCE's ongoing emphasis on institutional change and represent work done by later cohorts of ADVANCE Institutional Transformation institutions as well as

Table 6.3 Institutional and program design factors that support effective faculty mentoring as learned from the ADVANCE portfolio. (Adapted from DeAro et al., 2023)

Mentoring program design factors	Institutional context
Formal mentoring program(s) for everyone (not just a demographic subgroup or opt-in programs)	Clear, consistent, transparent policies, practices, and procedures related to evaluation, tenure, and promotion
Mentoring that is tailored for different career stages (early-, mid-, non-tenure track, etc.) and adjusted accordingly	Tracking and monitoring by administrators to include mentoring in workload distribution
Multiple points and modes of mentoring contact (circles, group mentoring, peer to peer, networking, etc.)	Inclusion of mentoring work in performance evaluations of faculty and department heads/chairs
Evidence-based training, support, and resources for mentors (and sometimes mentees)	Positive value given to mentoring in tenure and promotion decisions
Processes to deal with poor mentoring matches	

awardees of the other types of awards made by ADVANCE. We consider these projects "high-impact" because they have innovated strategies that have been adapted by other institutions of higher education. In our discussion of these projects, we draw on the typology by Laursen and Austin (see Table 6.2), including examples of strategies that *interrupt biased processes* and *reboot the institution.*

6.3.1 Interrupt Biased Processes

6.3.1.1 Advocates and Allies

One framework for culture and climate change can be found in the North Dakota State University's (NDSU) Advocates and Allies (A&A) program. The A&A program was initiated as part of NDSU's 2008 IT award (fourth cohort), *ADVANCE FORWARD—Focus on Resources for Women's Advancement, Recruitment/ Retention, and Development* (#0811239). A&A projects work to mitigate biased decision-making (especially on faculty hiring, tenure, and promotion committees where Allies serve) as well as improve the overall climate by enlisting all faculty in the work of creating equitable workplaces. NSF ADVANCE supported the expansion of the A&A model via an ADVANCE Partnership award that involved NDSU, the Ohio State University, the Rochester Institute of Technology, the University of North Texas, and the University of Wyoming (#1500604).

Data from a 2008 NDSU climate survey (NDSU 2010, p. 160) show that roughly three-quarters of women faculty respondents indicated that A&A had a very positive or somewhat positive impact on their experience at NDSU. In McGeorge and Bilen-Green's qualitative assessment of another A&A program, statements from women faculty members "provide some evidence that the Advocates and Allies program may be meeting its stated goal of intervening in ways that improve the campus climate for women faculty" (McGeorge & Bilen-Green, 2021, p. 45). The model has been adopted at over 40 different IHEs since 2008 (refer to Chapter 14 in this book for more details). The A&A model is an important component of the ADVANCE portfolio because it intentionally empowers and involves men in gender equity work. Those at NSF who developed ADVANCE understood that gender equity work is most frequently designed, implemented, and led by individuals who are part of the populations that experience the inequities. Aiming to shift that balance, the solicitation advised: "The burden of ADVANCE proposal development and implementation should not fall solely onto the populations that are intended to be served by this program. In order to address systemic issues all STEM faculty, administrators, and stakeholders likely need to be involved in the design and implementation and be potential participants in the project activities" (NSF, 2020).

6.3.1.2 Bystander Intervention Training

Bystander intervention training is a related strategy. In broad strokes, this involves training all faculty members to recognize bias when it happens and learn "… how to intervene in an impactful and safe manner" (UNH, n.d.) as articulated by the University of New Hampshire's Partnership project (*#1726351*). Numerous ADVANCE awards have adapted and innovated on this strategy, though the name varies between institutions, e.g., "active bystanders" at the University of Wisconsin Partnership project ADVANCEGeo (*#1725879 and #2204305,* UW, n.d.) and "bystander leadership" (FIU, n.d.-a), to name just two. These projects focus on empowering all faculty and administrators with the tools to create a culture of inclusion in their academic workplace and beyond.

Several of the ADVANCE-funded bystander intervention efforts have also included interactive theater. For example, the University of New Hampshire's virtual "IncludeU" Search Committee Training (*#1209189*) uses "interactive theater to highlight the best practices of the faculty search process and the impact of microaggressions and unconscious bias on a faculty search" (UNH, n.d.). Florida International University (FIU) used its Institutional Transformation grant's social science research project (*#1629889*) to inform both its Bystander Leadership program in general and its theater project in particular. The research studied "how a multiethnic cultural climate within FIU STEM departments affects the advancement of … Hispanic-American and African-American women faculty" and how "foreign-born … men's ethnic, racial and nationality-based gender stereotypes might create unique barriers for women of color in STEM" (FIU, n.d.-a, p. 21).

6.3.2 Reboot the Workplace

6.3.2.1 Alignment with Institutional Mission

Many ADVANCE projects were able to create sustainable, systemic change in institutions of higher education. Seattle University's *ADVANCE* project (#1629875, SU, n.d.-a) worked to align tenure and promotion criteria with the institution's mission, resulting in built-in sustainability. Seattle University (SU) is a predominantly undergraduate Jesuit institution whose values emphasize service and community engagement, but whose faculty promotion criteria had consistently undervalued this work. SU ADVANCE worked to transform both the policies and the institutional culture relating to service and community engagement, work that often falls disproportionately on faculty who are women, gender minorities, and/or racial minorities (O'Meara, 2017 and 2018). The project was accomplished in three phases. The first phase "involved 'strategic communication' and institutional ethnography with the aim of disrupting cultural ideals about what 'counts' in faculty advancement." In the second phase, the team developed and implemented new university standards for promotion to professor informed by the work from the first phase. Phase three is ongoing acculturation to the new policies on two fronts: "mentoring faculty to articulate intentional career paths that integrate historically undervalued activities and to make the case for the value of these activities under the new promotion guidelines" and "training with deans, department chairs, and faculty evaluation committees to raise awareness of the value of these contributions and provide guidance in how to recognize and assess faculty contributions holistically" (#1629875).

Outcomes from SU ADVANCE have been significant and positive. For the 20 years prior to the new policy, one to two women and faculty members of color petitioned successfully for promotion to full professor per year. Since the policy took effect, the numbers increased to 11 in 2021 and 9 in 2022, with 15 expected in 2023. This success has led to a subsequent ADVANCE Partnership award *Assessing Differential Faculty Contributions and Aligning Faculty Advancement with Mission Values* to Seattle University to work with two other Jesuit institutions, Gonzaga University and Saint Louis University, and to disseminate information to the 28 institutional members of the Association of Jesuit Colleges and Universities (#2204538, Veen, 2023). In line with the literature on systemic gender equity issues, SU and their partners have "… identified four general areas of under-valued faculty activities in which women and faculty of color are over-represented. These areas include: under-valued scholarship (including science education research, applied research, public-oriented research); community engagement; faculty administrative leadership; and institutional work to meet diversity and inclusion aims" (SU, n.d.-b).

6.3.2.2 Academic Workload

The Faculty Workload and Rewards Project of the University of Maryland (UMD), College Park, took on the persistent problem of systemic inequality in the distribution of service work among faculty. The program enables sustainable, systemic change with "a model that academic units can use to understand faculty workload inequality; diagnose its causes; and make concrete changes to policy and practice, with the overall goal of making the academic workplace a more just and equitable environment for all faculty" (UMD, n.d.-a). This ADVANCE Partnership project (#1463898) followed UMD's 2010 Institutional Transformation award titled *Towards an Institution for Inclusive Excellence* (#1008117).

The implementation of the Faculty Workload and Rewards model in an academic unit begins with a socialization exercise where units review the social science research on how faculty workload inequities arise and how they are detrimental to the IHE's diversity, equity, inclusion, and access (DEIA) efforts and departmental effectiveness. The process continues with the creation of a work activity dashboard, containing data on faculty workload that typically goes unrewarded and/or is "non-promotable" work. This workload dashboard informs the next step wherein units identify specific workload inequities and the pertinent policies and practices. The process concludes, like the Seattle University ADVANCE project, with another socialization exercise around "individual professional development related to aligning work time with work priorities." Key findings from this project show that unit work practices and conditions "that support equity are significant, positive predictors of workload satisfaction and perception of fairness" (UMD, n.d.-b, see also O'Meara et al., 2019).

6.4 Learning About the NSF ADVANCE Program

A more recent stream of research has begun to focus on what we have learned *about* the NSF ADVANCE Program. In a 2018 presentation, some results from a 2012 ADVANCE program level evaluation of the ADVANCE program were summarized (DeAro, 2018). The data indicated improvements in representation, with increases in the percent of women STEM faculty (from 16% to 24%) at the surveyed cohort one and two institutions (N=13 to 19 institutions) and within this growth was an improvement, albeit small, for women of color (from 2.4% to 3.8%). Institutions also reported increases in the percent of new hires who identify as women (from 25% to 35%) as well as in the percent of women in institutional leadership roles from 10% to 16%. It is important to note that these evaluation results are indicators of impact on the diversity and representation of STEM faculty; they do not capture the extent of the impact of these Institutional Transformation awards on systemic change, culture and climate change, and gender equity. One of the premises of the program is that even with full participation of all genders in STEM academic roles, there may still be gender-based inequities in the academic workplace. Examples of systemic equity issues that could persist include gender-based pay gaps, implicit biases and microaggressions, and unequal teaching, service, and mentoring workloads, especially the "non-promotable" workloads shouldered by women and racial and ethnic minority faculty (Baez, 2000, and Babcock, 2017).

A synthetic evaluation of ADVANCE is difficult especially since each institution is required to identify its own equity problems and to develop its own solutions and/or adapt solutions developed by others. Further, not all projects include the same equity issues. For example, one institution might focus on problems in promotion to full professor, and another might emphasize diversifying the ranks of leadership. Beyond this, definitions of success vary by the type and complexity of the identified problems, the difficulty in measuring impact on gender equity, and the timeline necessary to see measurable change. Longer timelines for the impact of some strategies are also complicated by changes in the institution's administrative leadership. Concrete policy changes might be made in 2 or 3 years (as in the case of the Seattle University ADVANCE IT grant discussed earlier), while efforts to change organizational climate and culture might take a decade or more to result in measurable changes among faculty.

Several recent studies have framed the ADVANCE program's impact in terms of the network of people, strategies, practices, publications, and ideas that have been created. This work traces the dissemination and diffusion of ADVANCE and learning associated with ADVANCE projects. This is very much in the spirit of the criterion of broader impacts at the heart of NSF merit review—funded projects are intended to have a reach and life beyond a single institution's award and to affect institutions that may never receive an ADVANCE award.

With funding from NSF's ADVANCE and Education Core Research (ECR) programs, Kathrin Zippel and her team at Northeastern University undertook a study of the ADVANCE network (#1836671 and #2000713). They identified 273 awards made by ADVANCE from 2001 to 2018 to 195 unique institutions using the NSF abstract database. They then traced ties between these institutions through award partnerships, advisory board links, institutional mobility of team members, and co-authorships through CV reviews and website searches. Zippel and her co-authors found an impact far beyond the 195 institutions directly funded; the ADVANCE network is more than three times larger than the universities that have received awards, containing 847 unique institutions (an interactive tool is available at https://zippelwebpage.github.io/zippelWebPage/).

Zippel and her team identified four kinds of institutional connections between these institutions. Links made through partnership awards accounted for 66 connections. The team identified 824 publications generated by ADVANCE, with 284 co-authorship relationships across institutions among them. There were 529 links created by external advisory board memberships and 826 created by job mobility between institutions. Among those in the latter category, some were ADVANCE PIs at multiple institutions—at least eight institutions received ADVANCE IT awards after an ADVANCE PI or co-PI from another university relocated there (Gold et al., 2022).

Research is also tracing the dissemination of specific ideas and practices through the mechanism of ADVANCE. For example, Zippel and her co-author (Nelson & Zippel, 2021) studied the diffusion of the idea of "implicit bias," which has been a core concept throughout the life of the program. The term first appeared in the ADVANCE lexicon in the very first cohort in 2001 in the Hunter College ADVANCE proposal, for which the PIs were social psychologists Virginia Valian and Vita Rabinowitz (#0123609). The term began to appear in NSF solicitations in 2007 after the publication of the National Academies of Science volume, *Beyond Bias and Barriers (National Research Council,* 2007), and subsequently appeared in dozens of ADVANCE projects. The term made its way into popular discourse in 2013 and is now very much a common understanding of how discrimination happens.

Nelson and Zippel (2021) believe that several factors contributed to the speed and spread of its adoption in higher education, among them the fact that the existence of bias is empirically demonstrable (e.g., Moss-Racusin et al., 2012) and hence more likely to appeal to STEM faculty who might otherwise be resistant to the legitimacy of social science. Perhaps more significantly, they argued that it is also "impartial"; theories of implicit bias contend that everyone—both majority and minority group members—behave in biased ways. Studies of the effectiveness of implicit bias training—both within and outside ADVANCE programs—have had mixed results. A long line of research by Frank Dobbin and Alexandra Kalev (e.g., Kalev et al., 2006; Dobbin & Kalev, 2022), some of it funded by NSF (#1444586), has found that implicit bias training does little to produce long-term attitudinal change or increase workplace diversity in industry settings.

One difference between these industry diversity training efforts and the ADVANCE grantees, however, is that in ADVANCE the training was typically done concurrently with structural changes. The structural changes to policy, practice, and procedure provided the guard rails to minimize the impact of implicit biases in decision-making. For example, the University of California Irvine IT project (#0123682) recruited and trained college equity advisors to serve on search committees outside of their own department. The University of Michigan, Ann Arbor, IT project (#0123571) implemented the practice of providing applicant pool data to departments when undertaking a search. This data serves to argue against the claim that there are no qualified candidates. Stewart and Valian (2018) summarize several actions that provosts, deans, chairs, and search committees can take to enhance their search and hiring processes (p. 192–194), such as but not limited to holding the department accountable to recruit applicants reflective of the diversity of doctoral recipients in their field(s).

Research is also beginning to assess specific tools for change developed in ADVANCE projects. One of the most commonly adopted of all ADVANCE strategies has been the STRIDE (Strategies and Tactics for Recruiting to Improve Diversity and Excellence) approach to faculty recruiting, which was developed in the very first cohort of ADVANCE Institutional Transformation awards at the University of Michigan, Ann Arbor (#0123571). Though Michigan's program contains many components, one of its core tools is the candidate rubric, which is intended to reduce bias by slowing down the evaluation process and facilitating comparison on objective criteria.

With funding from an ECR award (#1661306), Mary Blair Loy and her colleagues assessed the effectiveness of rubrics in reducing bias in searches. They focused on one engineering program over a period of 4 years. They found that the number of women hired increased after rubrics were adopted, though acknowledged that establishing causality was difficult because hiring is affected by a multitude of factors. However, they also found that, controlling for research productivity, seniority, and research impact (H-index), women candidates received lower scores on the rubric than men and that this bias disadvantaged junior women candidates the most (Blair-Loy et al., 2022). Recent work by Dawn Culpepper and colleagues, also funded by NSF, supports these findings (Culpepper et al., 2023). Neither of these studies rejects the use of rubrics, per se, but instead argues that they be used in conjunction with "strategic application in departmental meetings to counteract individual bias and check interactional bias during the discussion of candidates" (Blair-Loy et al., 2022, p. 37).

6.5 Learning by the NSF ADVANCE Program

Learning by the NSF ADVANCE program can be seen in changes to the program design over the 23 years of its existence. Overall, the evolution of the program reflects learning by NSF from each cohort of ADVANCE grantees and the relevant social, behavioral, and organizational research that continues to add to our understanding of systemic issues in higher education and gender equity in STEM.

6.5.1 Intersectionality

In a systematic review of the language in ADVANCE program solicitations and changes to the program design between 2001 and 2018, Laursen and De Welde (2019) provide an excellent summary of structural changes to the program and changes in the language in ADVANCE solicitations over the life of the program. Laursen finds that ADVANCE solicitations demonstrated "… growing attention to equity, rather than simply to representation, and to intersectionality—how gender, race, social class, and other identities intersect to produce disparate experiences and outcomes for individuals differently positioned in social systems" (Laursen & De Welde, 2019, p. 1).

One criticism of ADVANCE noted by Laursen and De Welde (2019) and made by others (Morimoto & Zajicek, 2014; Hunt et al., 2012; Cantor et al., 2014; Armstrong & Jovanovic, 2015, 2017) was that for the first 15 years, ADVANCE solicitations treated women in STEM faculty as an undifferentiated group. At the same time, a greater understanding was emerging about intersectionality and the need to specifically attend to the obstacles faced by women of color in STEM—e.g., greater isolation, higher service loads, and more frequent challenges to their legitimacy as scientists (Griffin et al., 2011, 2013; Dowd & Bensimon, 2015; O'Meara et al., 2017; Zambrana et al., 2015). In response, the program introduced a requirement that all ADVANCE proposals include attention to intersectionality in 2016. Introduced by Kimberlé Crenshaw (Crenshaw, 1989), the concept of intersectionality originally focused on the lack of recognition of the experiences of Black women in the judicial system resulting in injustice and marginalization in political discourse. As the term was adopted more widely in social science, the term began to be taken to mean the effects of multiple axes of inequality working at the same time to shape individual lives and institutional inequities (Cho et al. 2013, Collins & Bilge 2016, Corneille et al. 2019).

By introducing intersectionality as a requirement and an "Additional Solicitation Specific Review Criterion" in the 2016 solicitation, ADVANCE signaled a desire to require teams to think in more complicated, and more inclusive, ways about inequity. The solicitation required potential ADVANCE teams to integrate the concept into their approaches to addressing gender equity issues as well as their proposed solutions to effect systemic and culture change. In practical terms, proposers were asked to recognize the power dynamics created by multiple, intersecting identities such as disability status, gender identity, sexual orientation, foreign-born and foreign-trained status, and faculty appointment type. In 2023, 5 years after the introduction of the language on intersectionality, both NSF ADVANCE and the grantee community were still working out how to operationalize the concept of intersectionality in ADVANCE systemic change work. Here, we share some awards that have took on this challenge.

The University of Washington and NDSU earned an ADVANCE Partnership award to focus on the intersection of gender and disability called *AccessADVANCE* (#2017017). The *AccessADVANCE* project has created a community of practice around this topic as well as developed tools to help ADVANCE institutions to include disability in their gender equity work and become more welcoming for, and inclusive of, faculty with disabilities (University of Washington, n.d., and NDSU n.d.). This project was featured in an NSF ADVANCE 20th anniversary webinar (NSF, 2021) and has produced several tools and guides to support those interested in taking on this work at their institution available on their website www.washington.edu/doit/programs/advance.

An Institutional Transformation award made in 2023 to the University of Utah (UU) also exemplified the program's emphasis on intersectionality. This project (NSF Award #2319989, UU, 2023) included significant components designed to include the Lesbian, Gay, Bisexual, Transgender, Queer/Questioning, Intersex, and Agender/Asexual, inclusive (LGBTQIA+) community in their ADVANCE systemic change work building on prior research work (Patridge et al. 2014). The Utah project also focused on another marginalized group—non-tenure-track faculty. Although language allowing and encouraging the inclusion of non-tenure-track faculty in ADVANCE awards has been present in the solicitation since 2016 (NSF, 2016), not many projects have included this group. While the inclusion of non-tenure-track faculty is a change from ADVANCE's beginnings when it had an explicit focus on tenure-track faculty in STEM, the change reflects the reality that in many institutions non-tenure-track faculty are now the majority. In 2015, approximately 70% of the STEM and non-STEM faculty in the United States were non-tenure track. Non-tenure-track faculty teach about 50% of the courses at 4-year institutions and more at community colleges, and they are among the lowest-status and lowest-paid instructional faculty at most universities (GAO, 2017).

Another example illustrates an interesting adaptation of intersectionality: the UTRGV STRIVE Partnership project (#2204577) attended to equity issues identified by the proposers that involve mental health and well-being as well as race and gender (UTRGV, n.d.). The project focuses on training STEM department chairs at Hispanic-Serving Institutions (HSIs) on creating "… psychologically healthy workplaces for faculty, with specific emphasis on supporting Latina/x engagement and success." This project is an example of a project that targeted IHE leaders as important change agents in academia rather than burden faculty with additional programs that require time and effort—which we know is already in short supply. ADVANCE efforts that empower departmental leadership like the UTRGV project have the potential to develop equity-minded leaders who can positively impact the climate and culture of their departments now and wherever they may go over their administrative career.

6.6 ADVANCE Reach

Since 2001, about 70% of all ADVANCE program funds (~$405 M 2001–2023) have gone to support systemic change within single, nonprofit, institutions of higher education. Here, these types of projects are collectively referred to as Single Institution Systemic Change awards, or SISC. Although the underlying ADVANCE funding mechanism, total award amount, and project length differ, all of these projects attempted to systemically change some components of their own academic workplace. These IHEs can be thought of as "ADVANCE institutions." All ADVANCE Institutional Transformation (IT) and Adaptation awards and some Partnership for Adaptation, Implementation, and Dissemination (PAID) are SISC projects. With this definition in mind, there have been 133 ADVANCE SISC IHEs between 2001 and 2023 (see Table 6.4 in the Appendix). It is important to note that some of these projects ended more than 15 years ago and some were just starting their first year in 2024. These one hundred and thirty-three (133) ADVANCE SISC IHEs represent 3.9% of all US nonprofit IHEs (133/3376) (Carnegie, 2015). From the work done by Gold et al. (2022) discussed earlier, it is known that the people involved in implementing ADVANCE projects are mobile and take ADVANCE with them wherever they go. The ADVANCE program has invested about 30% of ADVANCE program funds in efforts to adapt and scale up the lessons learned from the SISC work. This portfolio includes a diverse set of projects; some examples have been discussed earlier including the NDSU Advocates and Allies Partnership grant and the UTRGV STRIVE Partnership grant (#1500604 and #2204577). The Partnership track is an important element of the ADVANCE program because it includes nonacademic organizations in the ADVANCE program systemic change efforts. We know that faculty work within a larger STEM enterprise than their own institution which may also have systemic issues (professional societies, publishers, accreditation entities, etc.). The Partnership track supported projects with significant regional or national reach and impact. Partnership awards may focus on one or more STEM discipline areas (engineering, geology, etc.) and/or a specific sector of academia (deans, department heads, etc.). Many of these awards reflected projects initiated by faculty at IHEs who serve in leadership roles in professional scientific organizations. About 54 ADVANCE Partnership awards made to IHEs did not focus on systemic change at the IHE; rather, these worked on a myriad of dissemination or scale-up efforts including training others on gender equity, supporting the adaptation of systemic change strategies, and working toward systemic changes within nonacademic organizations.

Many ADVANCE Partnership projects aimed to empower equity-minded academic leaders. One of the many lessons learned from SISC ADVANCE projects is the importance of leaders who can champion equity work in academia and set culture and climate expectations. An example is the ADVANCE work by ASEE (NSF Awards #1760002 and #2204520). The ASEE project "… provides tools, strategies, and resources to support academic engineering deans in identifying and removing barriers impeding recruitment, retention and advancement of women faculty from all backgrounds in engineering higher education" (ASEE, 2023). The 2022 ASEE project was designed to leverage the prior work and partner with ABET to "… diversify the accreditation decision-making bodies and increase leadership accountability …" among other components (#2204520). Given the role of ASEE, these two ADVANCE projects could catalyze important systemic change across the country in engineering schools.

One persistent challenge for the program is capturing and sharing the knowledge created from the ADVANCE program awards in a systemic and enduring way. Fortunately, ADVANCE grantees have published their work in various formats to generously share what they have learned from this work. A recent example is the open-access book "Uprooting Bias in the Academy, Lessons from the Field" with lessons learned and advice to others doing similar work from the University of California at Davis ADVANCE Institutional Transformation project edited by Bisson et al. (2022). Unfortunately, like this chapter, these are static resources and cannot be updated as new ideas challenge old ideas and innovations begin to show impact. The *ADVANCE Journal*, started by Oregon State University with their IT award (#1409171), was an effort to provide a forum for timely discussion and knowledge sharing. It is an open access online journal designed "...to publish peer-reviewed scholarship related to institutional transformation concerning inclusion, equity, and justice in higher education." (https://www.advancejournal.org/). The ADVANCE program also supported the development of the ADVANCE Resource Coordination Network (ARC Network), a partnership between NSF and the Women in Engineering Programs and Advocates Network (WEPAN).[1] From their website, "The ARC Network strives to advance equity nationally by facilitating the adoption and implementation of evidence-based systemic initiatives by institutions of higher education and other STEM organizations …" (see archived web site: https://web.archive.org/web/20250122192943/ https://www.equityinstem.org/). An important aspect of the ARC Network was that anyone interested in intersectional gender equity could join and attend the ADVANCE Equity in STEM Community Convening, participate in emerging research conferences and webinars, or apply to be a virtual visiting scholar. Most importantly, the ARC Network curated a library of over 9,000 resources related to equity in STEM in a Mendeley group library. Prior efforts to consolidate the ADVANCE resources on web sites such as the Virginia Tech ADVANCE Portal while effective for a time, only worked because of the dedicated time and effort of a few individuals that were barely compensated. Unfortunately, the grant to the ARC Network was terminated by the NSF in 2025, which ended, among many other important activities, the active curation and support of the

[1] The ARC Network original started in partnership with the Association for Women in STEM (AWIS) in 2017 and moved to WEPAN in 2021.

ADVANCE resource library. Although the ARC Network library remains available on Mendeley, it is not clear if group members will be able to managed the library to keep it up to date and to keep it open to new members.

The AAAS' SEA Change (STEMM Equity Achievement Change) effort, championed by Shirley Malcolm since 2019, is another type of effort to scale up lessions learned from ADVANCE. SEA Change has introduced awards for institutions of higher education that demonstrate a commitment to diversity, equity, inclusion, and access to incentivize science, technology, engineering, mathematics, and medicine (STEMM) culture change. SEA Change was first funded by NSF ADVANCE with a conference grant to the University of Massachusetts Lowell (#1632847) and later supported by all the NSF Directorates and the National Institutes of Health to develop the infrastructure needed to launch the effort nationally (#1841687). This work builds on one of the main lessons learned from ADVANCE, which is the importance of collecting and using data in diversity, equity, inclusion and access work. As mentioned earlier, ADVANCE grantees realized early on that their institutions were not collecting faculty workforce data in ways that were useful for identifying potential equity issues. In the early days of the SEA Change effort, Malcolm would often refer to SEA Change as "ADVANCE 2.0" because of how the project built on lessons learned from ADVANCE and because it had the potential to catalyze systemic change throughout higher education by rewarding institutions that collect and analyze data on their students, faculty, and staff. "SEA Change seeks to inspire, guide, and support voluntary transformation of colleges and universities so that the environment for research and education in science, technology, engineering, mathematics, and medicine (STEMM) is excellent, equitable, diverse, and inclusive" (seachange.aaas.org).

6.7 Diversity of ADVANCE Institution Types

Many changes to the types of awards supported by the program have resulted from a desire to diversify the pool of IHEs participating in the ADVANCE program as well as facilitate the transfer of ADVANCE lessons learned to other IHEs and organizations. The first two cohorts of ADVANCE Institutional Transformation (IT) projects were concentrated almost exclusively in research highest (R1) and higher (R2) IHEs (the exceptions were the University of Puerto Rico, Humacao, a primarily undergraduate IHE and a Hispanic-Serving Institution, and CUNY Hunter College, a master's degree-granting IHE). To encourage more diverse IHEs to participate, the ADVANCE program created the Partnerships for Adaptation, Implementation, and Dissemination track (PAID) in 2006. The Adaptation track was introduced in 2018, when the "A" was separated from the PAID mechanism, which then simply became the Partnership track. The Adaptation track was modeled on several awards that undertook smaller, shorter, SISC projects under the PAID track.

Figure 6.1 shows the number of ADVANCE awards made to IHEs to undertake Single Institution Systemic Change (SISC) over the last 23 years. The two different types of SISC in the graph are those made as Institutional Transformation (IT) awards (black) and those made under the PAID and the Adaptation program tracks (grey). Figure 6.1 illustrates the change in the program's investment, from large SISC awards ($3 M or more over 5 years) to more and smaller SISC awards ($1 M over 3 years). Though the underlying ADVANCE funding mechanism, total award amount, and project lengths differ, all of these projects attempted to systemically change components of the academic workplace to address gender inequities. This change from fewer large IT awards to more smaller Adaptation awards has resulted in an increase in the number of IHEs reached by the program to do SISC work (scale up), as well as resulted in more institutional type diversity in the ADVANCE portfolio. In just 5 years (2018–2023), the ADVANCE program increased the total number of IHEs undertaking SISC work by 34% (45 of the 133 SISC institutions have been funded under the Adaptation track).

Before the Adaptation track, the program introduced the Catalyst track to fund institutional self-assessment work to collect and analyze institutional data on STEM faculty. Proposing IHEs without the infrastructure to do data collection and analysis before submitting a proposal were at a disadvantage to IHEs that had the resources to do this work. Catalyst-type awards first formally appeared in 2008 as IT-Start,

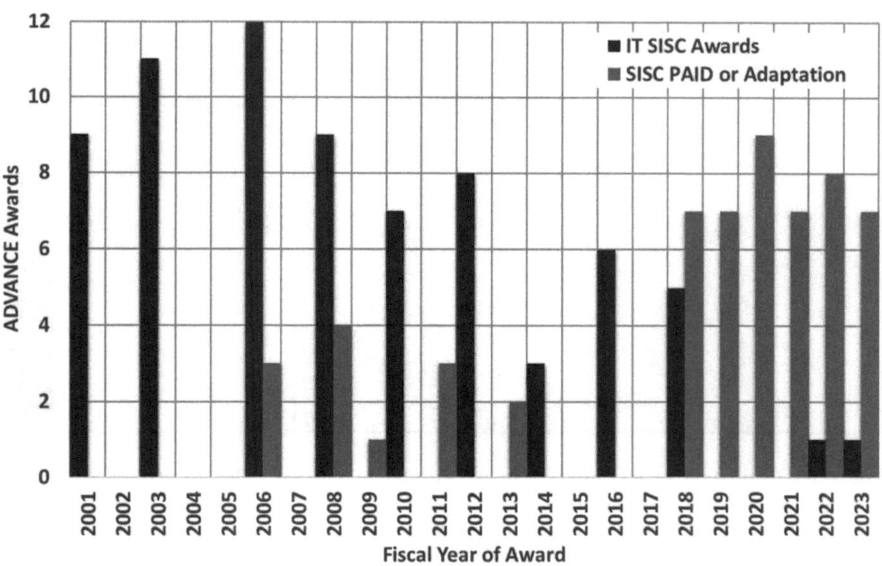

Fig. 6.1 Single Institution Systemic Change Awards by Mechanism over Time. The chart shows the change in numbers of awards made under the Institutional Transformation (2001–2023) SISC and the PAID (2006–2013) or Adaptation (2018–2023) SISC awards over time. The figure shows a transition, from more IT awards to more Adaptation awards in 2018. Importantly, the IT option remained in the program to support IHEs that proposed innovation in systemic change with two new IT awards made in 2022 and 2023

which then became IT-Catalyst and finally just Catalyst. One small award was made in 2003, which informed the Catalyst track. Since 2001, sixty-nine (69) of these small awards have been made to IHEs to undertake institutional self-assessment work in preparation for future systemic change efforts (these awards have ranged from $150 to $300 K over 2 years). There are indeed more diverse IHE types in the Catalyst cohort of ADVANCE awards as seen in Fig. 6.2. In addition, 21% of Catalyst IHEs are "inclusive" admissions institutions, while only 9% of the ADVANCE SISC IHEs are in this category (Carnegie, 2015).

Institutional self-assessment awards were started originally to create a pathway to the Institutional Transformation grant for under-resourced institutions and to diversify the IHE types participating in the program (DeAro et al., 2019). Twenty-six of the sixty-nine Catalyst IHEs were ongoing at the time of publication of this chapter, and would not have transitioned to additional ADVANCE funding. To date, 28% (12 of the 43 that ended their work) have gone on to undertake an SISC award via the Adaptation or IT tracks.[2] This transition rate from Catalyst to an ADVANCE

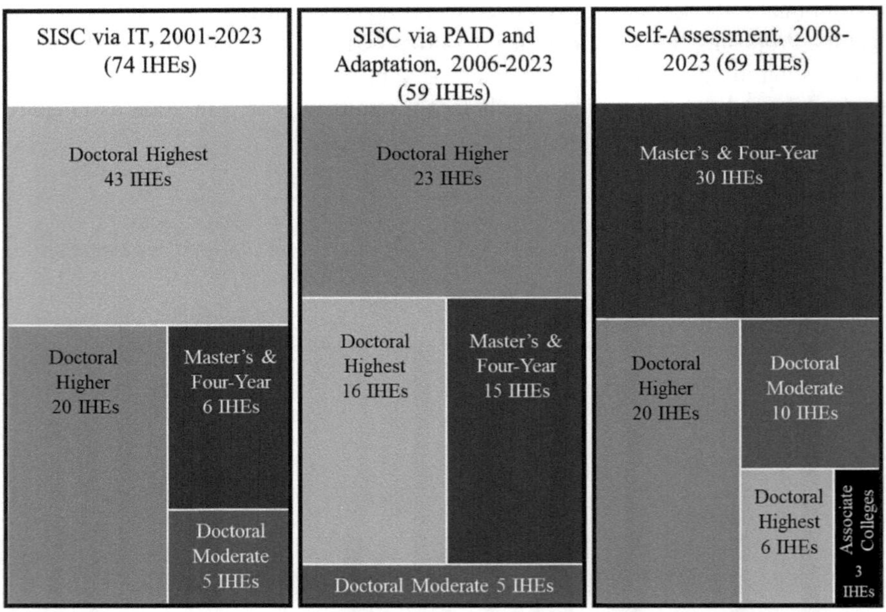

Fig. 6.2 Distribution of ADVANCE Single Institution Systemic Change and Catalyst Self-Assessment awards by type of IHE (2001–2023). The types of institutions with SISC projects change depending on the funding mechanism. A majority of IT awards have been made to Doctoral Highest (R1) and Doctoral Higher (R2) IHEs. PAID and Adaptation awards include more 4-year and master's-level institutions. The Catalyst (institutional self-assessment) projects have even more diversity in institutional types including associate-level colleges

[2] Note that these 12 IHEs are included in both the 133 SISC IHEs and the 69 Catalyst IHEs and so are double counted.

Adaptation or IT award may be good news in that many institutions that undertook the institutional self-assessment work may have found little to no gender equity issues or, more likely, they were able to make changes to their policies and procedures without additional federal grant funds. With this in mind in 2019, the Catalyst track was redesigned to catalyze change at the IHE, whether or not additional funds were sought, by requiring the development of a 5-year strategic plan for STEM faculty gender equity. An example of one of these 5-year strategic plans can be found on the Wichita State University ADVANCE Catalyst project (#1937921) website by Twomey et al. (n.d.).

Figure 6.2 visualizes the change in the types of IHEs that have SISC awards as a result of adding the PAID and Adaptation opportunities to the program. The IHEs that received Catalyst awards to undertake institutional assessment projects are even more diverse, with three associate degree-granting institutions (the first ever funded by ADVANCE) and five times as many master's and 4-year IHEs than in the IT track. While the program does not require an institution to start with the Catalyst, then the Adaptation, and then the IT, this three-step option was available to any IHE that had not already had an IT or Adaptation award. If an institution pursued this pathway, they could be working on systemic change for at least 10 years (2 years Catalyst + 3 years Adaptation + 5 years IT). This potential long pathway was purposeful, as one of the lessons learned from the early cohorts of IT awardees is that systemic change, especially culture and climate change, takes time—usually more than the 5 years of an IT award.

6.8 Conclusion

Since 2001, the NSF ADVANCE program learned, grew, and changed. ADVANCE grantees have produced new knowledge about addressing gender-based systemic inequities in higher education settings for STEM faculty. The ADVANCE program itself has contributed to learning and new knowledge about how to implement systemic changes with an equity lens. The program design and emphasis have changed over time as the program has learned from its grantees about the *process* of addressing systemic inequities, and the *strategies* and *programs* that have the most potential for achieving long-term change in institutions of higher education.

The ADVANCE community of grantees that developed over 25 years is one of the best unintended consequences of the ADVANCE program. This community of change agents and researchers support and celebrate each other and are generous in sharing what they know with anyone interested in gender equity and systemic change work. The community created the ADVANCE Implementation Mentors (AIM) Network to create a mechanism to support individuals managing ADVANCE projects in 2010 as a way to share best practices and lessons learned (https://aimadvancenetwork.org/). Fortunately, both the AIM Network and the *ADVANCE Journal* remain active despite the archiving of the program in 2025. The number of person hours (and sweat and tears) dedicated to researching, proposing, implementing, and evaluating ADVANCE projects should be considered a priceless gift to the country and to future STEM students and faculty. The authors wish to thank the many sung, and unsung, ADVANCE contributors over the years and welcome collaborations with all who seek to contribute to this work via an NSF program or other venture.

Finally, while there is no indication that the ADVANCE program or something similar will be available from NSF at any time in the future, the contributions of the program over the last 25 years are undeniable and have made a mark on higher education and thousands of faculty and administrators. The authors maintain that more equitable academic environments will attract and retain more diverse faculty, and this diversity is fundamentally linked to innovation and excellence in scientific discovery and the US economy and national security. As debates over diversity, equity, inclusion, and access in higher education in the United States have intensified in recent years, the goals espoused by ADVANCE have never been more important.

Disclaimer The views expressed in this chapter do not necessarily reflect the views of the U.S. National Science Foundation or the United States.

Appendix: Single Institution Systemic Change (SISC) NSF ADVANCE Awards 2001–2025 (Table 6.4)

Note: The NSF ADVANCE Program was archived in 2025 and most on-going ADVANCE awards were terminated by the NSF.

Table 6.4 NSF ADVANCE awards 2001–2025

Institutional Transformation (IT) awards

IT Cohort 1 (2001)[a]

1. Hunter College, CUNY (NY)
2. University of Colorado, Boulder (CO)
3. Georgia Institute of Technology (GA)
4. University of Michigan, Ann Arbor (MI)
5. New Mexico State University (NM)
6. University of Puerto Rico, Humacao (PR)
7. University of California, Irvine (CA)
8. University of Washington (WA)
9. University of Wisconsin, Madison (WI)

IT Cohort 2 (2003)[a]

1. Case Western Reserve University (OH)
2. University of Montana (MT)
3. Columbia University (NY)
4. University of Rhode Island (RI)
5. University of Alabama, Birmingham (AL)
6. University of Texas, El Paso (TX)
7. Kansas State University (KS).
8. Utah State University (UT)
9. University of Maryland, Baltimore County (MD)
10. Virginia Polytechnic Institute (VA)

IT Cohort 3 (2006)

1. Brown University (RI)
2. Cal State Polytechnic University, Pomona (CA)
3. Cornell University (NY)
4. Iowa State University (IA)
5. Marshall University (WV)
6. New Jersey Institute of Technology (NJ)
7. Rensselaer Polytechnic Institute (NY)
8. University of Arizona (AZ)
9. University of Illinois, Chicago (IL)[b]
10. University of North Carolina, Charlotte (NC)
11. William Marsh Rice University (TX)
12. Duke University (NC)[c]
13. University of Maryland Eastern Shore (MD)[c]

IT Cohort 4 (2008)

1. Michigan State University (MI)
2. North Dakota State University, Fargo (ND)
3. Northeastern University (MA)
4. Ohio State University (OH)
5. Purdue University (IN)
6. Rutgers University (NJ) all three campuses—Camden, Newark, and New Brunswick
7. University of Nebraska, Lincoln (NE)
8. Washington State University (WA)
9. Wright State University (OH)

IT Cohort 5 (2010)

1. West Virginia University (WV)
2. University of Maryland, College Park (MD)
3. Lehigh University (PA)
4. Texas A&M University (TX)
5. University of Maine (MN)
6. Syracuse University (NY)
7. Jackson State University (MS)

IT Cohort 6 (2012)

1. Montana State University (MT)
2. Howard University (DC)
3. Rochester Institute of Technology (NY)[b]
4. University of Cincinnati (OH)
5. University of New Hampshire (NH)
6. University of Virginia (VA)
7. University of Texas RGV (TX)
8. University of CA Davis (CA)

IT Cohort 7 (2014)

1. Oregon State University (OR)
2. University of Delaware (DE)
3. North Carolina Agricultural and Technical University (NC)
4. University of Houston (TX)

IT Cohort 8 (2016)

1. University of New Mexico (NM)
2. Clemson University (SC)
3. Florida International University (FL)
4. Seattle University (WA)
5. University of Massachusetts Lowell (MA)

IT awards made after IT-Preliminary proposal required in 2016:

IT Cohort 9 (2018)

1. Arizona State University (AZ)
2. Florida A&M University (FL)
3. University of Massachusetts Amherst (MA)
4. Villanova University (PA)
5. Virginia Commonwealth University (VA)

Post-Cohort IT Awards

1. University of Oklahoma (OK) (2022)
2. University of Utah (UT) (2023)
3. Washington University (MO) (2024)

continued

Table 6.4 (continued)

Institutional Transformation (IT) awards	
PAID SISC Awards	
1. Auburn University (2006) 2. Grand Valley State University (2006) 3. Idaho State University (2006) 4. Boston University (2008) 5. North Carolina State University at Raleigh (2008) 6. Oklahoma State University (2008) 7. University of Miami (2008)	8. Louisiana Tech University (2009) 9. College of New Jersey (2009) 10. Oakland University (2011) 11. University of West Florida (2011) 12. University of California-Riverside (2011) 13. Medical University of South Carolina (2013) 14. Stevens Institute of Technology (2013)
Adaptation SISC Awards	
Adaptation Cohort 1 (2018) 1. Bowling Green State University—Main Campus 2. California State University-San Bernardino 3. Michigan Technological University 4. University of Tennessee-Knoxville 5. University of North Carolina at Chapel Hill 6. Worcester Polytechnic Institute 7. Xavier University **Adaptation Cohort 2 (2019)** 1. Clarkson University 2. Indiana University–Purdue University, Indianapolis 3. Marquette University 4. Murray State University[b] 5. Southern Illinois University Edwardsville[b] 6. University of Louisville 7. University of North Carolina at Greensboro **Adaptation Cohort 3 (2020)** 1. Appalachian State University 2. Cleveland State University 3. East Carolina University 4. San Francisco State University[b] 5. Texas Tech University 6. University of Arkansas 7. University of Memphis 8. University of Missouri-St. Louis 9. Wayne State University **Adaptation Cohort 4 (2021)** 1. Colorado State University-Fort Collins 2. Loyola University Chicago	8. South Dakota School of Mines and Technology[b] 9. SUNY at Albany 10. University of Texas at Dallas 11. University of Colorado, Colorado Springs 12. University of Denver **Adaptation Cohort 5 (2022)** 1. American University[b] 2. California State University, Long Beach 3. CUNY City College[b] 4. Florida Atlantic University[b] 5. Georgia State University 6. Missouri University of Science and Technology 7. Nevada State College 8. University of Wisconsin-Oshkosh **Adaptation Cohort 6 (2023)** 1. Bradley University 2. Central Michigan University[b] 3. Florida State University 4. New York University 5. Old Dominion University 6. University of Texas at San Antonio[b] 7. University of Massachusetts-Boston[b] **Adaptation Cohort 7 (2024)** 1. University of Illinois at Urbana-Champaign 2. Missouri State University[b] 3. University of North Texas 4. Wichita State University[b] 5. Eastern Washington University[b]

[a] These IT awardees were required to provide 20% cost sharing, which increased their awards to about $4.3 M each

[b] IHE had a prior ADVANCE Catalyst grant (or similar) to undertake institutional self-assessment work

[c] These awards were shorter and smaller than the typical IT—similar to the Adaptation opportunity

References

American Society for Engineering Education. (2023). ASEE engineering deans gender equity (EDGE) initiative. Retrieved from https://edge.asee.org/

Armstrong, M. A., & Jovanovic, J. (2015). Starting at the crossroads: Intersectional approaches to institutionally supporting underrepresented minority women stem faculty. *Journal of Women and Minorities in Science and Engineering, 21*(2), 141–157. https://doi.org/10.1615/JWomenMinorScienEng.2015011275

Armstrong, M. A., & Jovanovic, J. (2017). The intersectional matrix: Rethinking institutional change for URM women in STEM. *Journal of Diversity in Higher Education, 10*(3), 216–231. https://doi.org/10.1037/dhe0000021

Babcock, L., Recalde M. P., Vesterlund, L., and Weingart, L. (2017). Gender Differences in Accepting and Receiving Requests for Tasks with Low Promotability. *American Economic Review 107*(3), 714–47. https://doi.org/10.1257/aer.20141734

Baez, B. (2000). Race-related service and faculty of color: Conceptualizing critical agency in academe. Higher Education 39, 363–391. https://doi.org/10.1023/A:1003972214943

Bilimoria, D., & Liang, X. (2012). *Gender equity in science and engineering: Advancing change in higher education.* Routledge.

Bisson, L. F., Grindstaff, L., Brazil-Cruz, L., & Barbu, S. (2022). *Uprooting bias in the academy, lessons from the field.* Springer. https://doi.org/10.1007/978-3-030-85668-7

Blair-Loy, M., Mayorova, O. V., Cosman, P. C., & Fraley, S. I. (2022). Can rubrics combat gender bias in faculty hiring? *Science, 377*(6601), 35–37. https://doi.org/10.1126/science.abm2329

Cantor, N., Mack, K., McDermott, P., & Taylor, O. L. (2014). If not now, when? The promise of STEM intersectionality in the twenty-first century. *Peer Review, 16*(2), 29–31.

Carnegie Classification of Institutions of Higher Education®. (2015). Retrieved January 19, 2024, from https://carnegieclassifications.acenet.edu/

Cho, S., Crenshaw, K. W., & McCall, L. (2013). Toward a field of intersectionality studies: Theory, applications, and praxis. *Signs: Journal of Women in Culture and Society, 38*(4), 785–810. https://doi.org/10.1086/669608

Corneille, M., Lee, A., Allen, S., Cannady, J., & Guess, A. (2019). Barriers to the advancement of women of color faculty in STEM: The need for promoting equity using an intersectional framework. *Equality, Diversity and Inclusion: An International Journal, 38*(3), 328–348. https://doi.org/10.1108/EDI-09-2017-0199

Crenshaw, K. (1989). Demarginalizing the intersection of race and sex: A black feminist critique of antidiscrimination doctrine, feminist theory and antiracist policies. University of Chicago Legal Forum, Vol. 1989, Article 8, 139–67. http://chicagounbound.uchicago.edu/uclf/vol1989/iss1/8/

Culpepper, D., White-Lewis, D., O'Meara, K., Templeton, L., & Anderson, J. (2023). Do rubrics live up to their promise? Examining how rubrics mitigate bias in faculty hiring. *Journal of Higher Education, 94*(7), 823–850. https://doi.org/10.1080/00221546.2023.2168411

DeAro, J., Bird, S., & Ryan, S. M. (2019). NSF ADVANCE and gender equity past, present and future of systemic institutional transformation strategies. *Equality, Diversity and Inclusion: An International Journal, 38*(2, SI), 131–139. https://doi.org/10.1108/EDI-09-2017-0188

DeAro, J. A. (2018). Case Western Reserve University IDEAL-N Annual Plenary 2018. Retrieved from https://case.edu/ideal-n/presentations-and-publications/annual-plenary-2018-presentations

DeAro, J., Amador, J., Byers-Winston, A., Pfund, C., & Salt, K. (2023). *How does mentoring impact diversity in the biomedical and behavioral research workforce? | COSWD* (Workforce Diversity Seminar Series). National Institutes of Health. Retrieved from https://diversity.nih.gov/disseminate/swd-seminar-series/how-does-mentoring-impact-diversity-biomedical-and-behavioral

Dobbin, F., & Kalev, A. (2022). *Getting to diversity: What works and what doesn't.* The Belknap Press of Harvard University Press.

Dowd, A. C., & Bensimon, E. M. (2015). *Engaging the "race question": Accountability and equity in U.S. higher education*. Teachers College, Columbia University.

Dyer, R. A., & Montelone, B. A. (2007). An institutional approach to establishing professional connections. In *Transforming science and engineering: Advancing academic women* (pp. 48–61). University of Michigan Press.

Florida International University (FIU). (n.d.-a). Florida International University—Year 6 Annual Report. Retrieved on December 1, 2023, from https://awed.fiu.edu/about/impact-and-reporting/year-6-advance-report-final.pdf

Florida International University (FIU). (n.d.-b). Office to advance women, equity and diversity. Retrieved from https://awed.fiu.edu/. Accessed 28 Jan 2024

Gold, J. R., Gates, A. J., Haque, S. A., Melson, M. C., Nelson, L. K., & Zippel, K. (2022). The NSF ADVANCE Network of Organizations. *ADVANCE J., 3*(1). https://doi.org/10.5399/osu/ADVJRNL.3.1.3

Government Accountability Office. (2017). Contingent workforce size, characteristics, compensation, and work experiences of adjunct and other non-tenure-track faculty. (GAO 18–49). Retrieved from https://www.gao.gov/products/gao-18-49

Griffin, K. A., Bennett, J. C., & Harris, J. (2013). Marginalizing merit? Gender differences in black faculty D/discourses on tenure, advancement, and professional success. *The Review of Higher Education, 36*(4), 489–512. https://doi.org/10.1353/rhe.2013.0040

Griffin, K. A., Pifer, M. J., Humphrey, J. R., & Hazelwood, A. M. (2011). (Re)defining departure: Exploring black professors' experiences with and responses to racism and racial climate. *American Journal of Education, 117*(4), 495–526. https://doi.org/10.1086/660756

Hill Collins, P., & Bilge, S. (2016). *Intersectionality*. Wiley.

Hunt, V. H., Morimoto, S., Zajicek, A., & Lisnic, R. (2012). Intersectionality and dismantling institutional privilege: The case of the NSF ADVANCE program. *Race Gender Class*, 266–290.

Jackson State University (JSU). (n.d.-a). Summer writing retreat. Retrieved on January 27, 2024 from https://www.jsums.edu/jsuadvance/programs/summer-writing-retreats/

Jackson State University (JSU). (n.d.-b). Policy review, modification and adoption. Retrieved on January 27, 2024 from https://www.jsums.edu/jsuadvance/programs/policy-review-modification-and-adoption/

Kalev, A., Dobbin, F., & Kelly, E. (2006). Best practices or best guesses? Assessing the efficacy of corporate affirmative action and diversity policies. *American Sociological Review, 71*(4), 589–617. https://doi.org/10.1177/000312240607100404

Laursen, S., & Austin, A. E. (2020). *Building gender equity in the academy: Institutional strategies for change*. Johns Hopkins University Press.

Laursen, S. L., & De Welde, K. (2019). The changer and the changed: Evolving theories and practices of change in ADVANCE calls for institutional transformation. *Equality, Diversity and Inclusion: An International Journal, 38*(2), 140–159. https://doi.org/10.1108/EDI-09-2017-0192

McGeorge, C. R., & Bilen-Green, C. (2021). Engaging men as allies for gender equity in higher education: An exploration of an advocates and allies program. *Journal of Women and Minorities in Science and Engineering, 27*(2), 25–59. https://doi.org/10.1615/JWomenMinorScienEng.2020031266

Michigan Tech Early Career Management | Faculty Development | Michigan Tech. (n.d.). Michigan Technological University Retrieved January 19, 2024, from https://www.mtu.edu/faculty-development/mentoring/ecm/

Morimoto, S. A., & Zajicek, A. (2014). Dismantling the 'master's house': Feminist reflections on institutional transformation. *Critical Sociology, 40*(1), 135–150. https://doi.org/10.1177/0896920512460063

Moss-Racusin, C. A., Dovidio, J. F., Brescoll, V. L., Graham, M. J., & Handelsman, J. (2012). Science faculty's subtle gender biases favor male students. *Proceedings of the National Academy of Sciences, 109*(41), 16474–16479. https://doi.org/10.1073/pnas.1211286109

National Research Council. (2007). *Beyond bias and barriers: Fulfilling the potential of women in academic science and engineering*. The National Academies Press. Retrieved from http://www.nap.edu/catalog/11741/beyond-bias-and-barriers-fulfilling-the-potential-of-women-in

National Science Foundation. (2016). ADVANCE: Increasing the participation and advancement of women in academic science and engineering careers (ADVANCE) (NSF16-594) | NSF—National Science Foundation. Retrieved from https://www.nsf.gov/pubs/2016/nsf16594/nsf16594.htm

National Science Foundation. (2020). ADVANCE: Organizational change for gender equity in STEM academic professions (ADVANCE) (NSF20-554). Retrieved from https://new.nsf.gov/funding/opportunities/advance-organizational-change-gender-equity-stem/nsf20-554/solicitation

National Science Foundation. (March 3, 2021). Recorded 20th Anniversary ADVANCE Presentations & Panels | NSF—National Science Foundation. Retrieved from https://web.archive.org/web/20230601032113/https://new.nsf.gov/events/recorded-20th-anniversary-advance-presentations

Nelson, L. K., & Zippel, K. (2021). From theory to practice and back: How the concept of implicit bias was implemented in academe, and what this means for gender theories of organizational change. *Gender and Society, 35*(3), 330–357. https://doi.org/10.1177/08912432211000335

North Dakota State University (NDSU). (2010). Faculty work/life balance: Results of the 2008 NDSU FORWARD survey. Retrieved from https://www.ndsu.edu/fileadmin/forward/documents/Work_Life_Final_Report_October_2010_with_cover_2_.pdf

NDSU. (n.d.). Women with Disabilities Task Force, Retrieved January 19, 2024, from https://www.ndsu.edu/forward/institutional_transformation_project/archives/advance_forward_initiatives/women_with_disabilities_task_force/

O'Meara, K., Jaeger, A., Misra, J., Lennartz, C., & Kuvaeva, A. (2018). Undoing disparities in faculty workloads: A randomized trial experiment. *PLoS One, 13*(12), e0207316. https://doi.org/10.1371/journal.pone.0207316

O'Meara, K., Kuvaeva, A., Nyunt, G., Waugaman, C., & Jackson, R. (2017). Asked more often: Gender differences in faculty workload in research universities and the work interactions that shape them. *American Educational Research Journal, 54*(6), 1154–1186. https://doi.org/10.3102/0002831217716767

O'Meara, K., Lennartz, C. J., Kuvaeva, A., Jaeger, A., & Misra, J. (2019). Department conditions and practices associated with faculty workload satisfaction and perceptions of equity. *The Journal of Higher Education, 90*(5), 744–772. https://doi.org/10.1080/00221546.2019.1584025

Patridge, E. V., Barthelemy, R. S., & Rankin, S. R. (2014). Factors impacting the academic climate for LGBQ stem faculty. *Journal of Women and Minorities in Science and Engineering, 20*(1), 75–98. https://doi.org/10.1615/JWomenMinorScienEng.2014007429

Rosser, S. V. (2001). Balancing: Survey of fiscal year 1997, 1998, and 1999 POWRE awardees. *Journal of Women and Minorities in Science and Engineering, 7*(1), 1–11.

Rosser, S. V. (2004). Using POWRE to ADVANCE: Institutional barriers identified by women scientists and engineers. *NWSA Journal, 16*(1), 50–78. https://doi.org/10.2979/NWS.2004.16.1.50

Rosser, S. V. (2012). *Breaking into the Lab. Engineering progress for women in science*. New York University Press.

Seattle University (SU). (n.d.-a). Seattle University ADVANCE. Retrieved January 19, 2024, from https://www.seattleu.edu/advance/

Seattle University (SU). (n.d.-b). *AJCU Partnerships | Seattle University ADVANCE Program*. Seattle University. Retrieved January 19, 2024, from. https://www.seattleu.edu/advance/ajcu-partnerships/

Spoon, K., LaBerge, N., Wapman, K. H., Zhang, S., Morgan, A. C., Galesic, M., Fosdick, B. K., Larremore, D. B., & Clauset, A. (2023). Gender and retention patterns among U.S. faculty. *Science Advances, 9*(42), eadi2205. https://doi.org/10.1126/sciadv.adi2205

Stewart, A., Malley, J., & LaVaque-Manty, D. (2007). *Transforming science and engineering: Advancing academic women*. University of Michigan Press. https://doi.org/10.3998/mpub.178866

Stewart, A. J., Malley, J. E., & Linderman, J. J. (2021). Launching new faculty careers: Building a strong foundation for a diverse faculty. *Journal of Women and Minorities in Science and Engineering, 27*(1), 85–106. https://doi.org/10.1615/JWomenMinorScienEng.2020035401

Stewart, A. J., & Valian, V. (2018). *An inclusive academy: Achieving diversity and excellence*. The MIT Press.

Thomas, N., Bystydzienski, J., & Desai, A. (2015). Changing institutional culture through peer mentoring of women STEM faculty. *Innovative Higher Education, 40*(2), 143–157. https://doi.org/10.1007/s10755-014-9300-9

Twomey, J. Glenmaye, L., Lewis, R. Beck, M. Eichhorn, D., Franks, T., Griffith, J., & Markova, G. (n.d.). Five year strategic plan to remove systemic barriers that have led to the paucity of female and minority STEM faculty at all academic tans and their advancement into leadership positions within WSU. Retrieved on January 27, 2024 from https://www.wichita.edu/academics/academic_affairs/advance/_documents/Strategic_plan_9-21-21.pdf

University of Maryland (UMD). (n.d.-a). ADVANCE: Investing in faculty success and inclusive work environments. Retrieved January 19, 2024, from https://advance.umd.edu/

University of Maryland (UMD). (n.d.-b). The faculty workload and rewards project. Retrieved January 19, 2024, from https://advance.umd.edu/fwrp/about/overview

University of New Hampshire (UNH). (n.d.). ADVANCE partnership grant. UNH ADVANCE. Retrieved January 19, 2024, from https://www.unh.edu/professional-success/unh-advance/about/unh-advance-partnership-grant

University of Texas Rio Grande Valley. (n.d.). UTRGV ADVANCE: Gender diversity matters THRIVE partnership. Retrieved from https://www.utrgv.edu/advance/initiatives-events/thrive-partnership/index.htm

University of Texas Rio Grande Valley (UTRGV). (n.d.). Retrieved from https://www.utrgv.edu/advance/initiatives-events/thrive-partnership/index.htm

University of Utah (UU). (October 5, 2023). U looks to ADVANCE women faculty in STEM—@theU. Retrieved from https://attheu.utah.edu/facultystaff/u-to-develop-policies-and-practices-to-support-women-faculty-in-stem-with-a-focus-on-lgbtq-and-women-of-color/

University of Washington ADVANCE. (n.d.). AccessADVANCE | DO-IT. Retrieved January 19, 2024, from https://www.washington.edu/doit/project/accessadvance

University of Wisconsin. (n.d.). ADVANCEGeo workshops. Workshops. Retrieved January 19, 2024, from https://serc.carleton.edu/advancegeo/workshops/index.html

Valian, V. (1998). *Why so slow? The advancement of women*. The MIT Press.

Veen, L. V. (2023). SU, Gonzaga and Saint Louis U Awarded Joint $1 million NSF Grant, SeattleU the Newsroom. Retrieved from https://www.seattleu.edu/newsroom/stories/2023/su-gonzaga-and-saint-louis-u-awarded-joint-1-million-nsf-grant-1.html

Zambrana, R. E., Ray, R., Espino, M. M., Castro, C., Douthirt Cohen, B., & Eliason, J. (2015). "Don't leave us behind": The importance of mentoring for underrepresented minority faculty. *American Educational Research Journal, 52*(1), 40–72. https://doi.org/10.3102/0002831214563063

ADVANCE Awards Mentioned in the Text

#0123571—ADVANCE Institutional Transformation Award. (2001). University of Michigan. Retrieved from https://www.nsf.gov/awardsearch/showAward?AWD_ID=0123571

#0123609—ADVANCE Institutional Transformation Award. (2002). Hunter College—CUNY. Retrieved from https://www.nsf.gov/awardsearch/showAward?AWD_ID=0123609&HistoricalAwards=false

#0244984—ADVANCE Institutional Transformation Award: Institutional Transformation at Kansas State University. (2003). Kansas State University. Retrieved from https://www.nsf.gov/awardsearch/showAward?AWD_ID=0244984&HistoricalAwards=false

#1008117—University of Maryland: Towards an Institution for Inclusive Excellence (UM=TI^2E). (2010). University of Maryland. Retrieved from https://www.nsf.gov/awardsearch/showAward?AWD_ID=1008117

#1209189—UNH UNBIASED: Leadership Development and Policy Change to Promote Institutional Transformation. (2012). University of New Hampshire. Retrieved from https://www.nsf.gov/awardsearch/showAward?AWD_ID=1209189

#1444586—Effects of ADVANCE in the STEM Disciplines: Faculty Diversity, Women in Leadership, and Institutional Transformation. (2014). Harvard University. Retrieved from https://www.nsf.gov/awardsearch/showAward?AWD_ID=1444586&HistoricalAwards=false

#1500604—Engaging Male Colleagues as Advocates and Allies for the Advancement of Women Faculty. (2015). North Dakota State University. Retrieved from https://www.nsf.gov/awardsearch/showAward?AWD_ID=1500604

#1629875—ADVANCE Institutional Transformation at Seattle University. (2016). Seattle University. Retrieved from https://www.nsf.gov/awardsearch/showAward?AWD_ID=1629875

#1629889—ADVANCE Institutional Transformation at Florida International University. (2016). Florida International University. Retrieved from https://www.nsf.gov/awardsearch/showAward?AWD_ID=1629889

#1725879—ADVANCE Partnership: From the Classroom to the Field: Improving the Workplace in the Geosciences. (2017). University of Wisconsin-Madison. Retrieved from https://www.nsf.gov/awardsearch/showAward?AWD_ID=1725879

#1760002—*ADVANCE Adaptation: ASEE Engineering Deans Gender Equity (EDGE) Initiative (2018)*. American Society for Engineering Education. Retrieved from https://www.nsf.gov/awardsearch/showAward?AWD_ID=1760002&HistoricalAwards=false

#1726351—ADVANCE Partnership: Faculty Intervention Guide and Decision Tool for Improving the Academic Workplace. (2017). University of New Hampshire. Retrieved from https://www.nsf.gov/awardsearch/showAward?AWD_ID=1726351

#1937921—ADVANCE Catalyst (n.d.)

#2017017—ADVANCE Partnership: AccessADVANCE. (2021). University of Washington. Retrieved from https://www.nsf.gov/awardsearch/showAward?AWD_ID=2017017

#2204305—ADVANCE Partnership: Empowering scientists to transform workplace climate through the ADVANCEGeo community-based intervention program. (2022). University of Wisconsin-Madison. Retrieved from https://www.nsf.gov/awardsearch/showAward?AWD_ID=2204305

#2204520—ADVANCE Partnership: Engineering Deans' Gender Equity Initiative: Aligning Systems to Ensure Inclusion and Equity in Advancement of Faculty. (2022). American Society for Engineering Education. Retrieved from https://www.nsf.gov/awardsearch/showAward?AWD_ID=2204520&HistoricalAwards=false

#2204538—ADVANCE Partnership: Assessing Differential Faculty Contributions and Aligning Faculty Advancement with Mission Values. (2022). University of Seattle. Retrieved from https://www.nsf.gov/awardsearch/showAward?AWD_ID=2204538

#2204577—ADVANCE Partnership: Transforming HSI STEM Departmental Climate for Faculty Success. (2022). University of Texas Rio Grande Valley Retrieved from https://www.nsf.gov/awardsearch/showAward?AWD_ID=2204577

#2319989—ADVANCE IT: University of Utah Institutional Change Studios' Systemic Institutional Change Through Problem Driven Iterative Adaptation. (2023). University of Utah. Retrieved from https://www.nsf.gov/awardsearch/showAward?AWD_ID=2319989

Chapter 7
Foreign-Born Women Faculty: Increasing in Academia

Sue V. Rosser

7.1 Introduction and Background

During the last two decades, the number of international or foreign-born/foreign-trained (FB/FT) faculty has increased dramatically in many science, technology, engineering, and mathematics (STEM) academic departments, so that they now constitute more than a majority of faculty in most engineering, computer science, physical sciences, and mathematics departments in the USA (NSB-NSF, 2021). NSB-NSF (2021) documents that the number of foreign students with doctoral degrees in STEM has been consistently increasing for almost two decades, and hiring FB women in academia has also increased simultaneously. For example, in engineering, 58.5% of PhD degrees are earned by FB students, and 54% of the women receiving PhDs in engineering are FB students (ASEE, 2023). Since many of the FB students seek to remain in the USA, it is not surprising that FB faculty now constitute more than 70% of faculty in many engineering departments.

Concurrently, universities have attempted to diversify their faculties, including those in STEM, with regard to gender and US-born/US-trained (USB/UST) underrepresented minority (URM) faculty, partially in response to the imperative to have the faculty mirror the increasingly diverse student body and US population. A burgeoning body of literature explores issues of surrounding experiences of gender bias, discrimination, career progression, and retention faced by STEM faculty, and increasing numbers of studies explore the experiences of racial/ethnic bias and discrimination of STEM faculty. The URM categories used by the NSF and U.S. Census Bureau do not always correspond with how FB/FT faculty identify, raising questions as to the extent that studies of URM faculty also apply to FB/FT faculty of color.

S. V. Rosser (✉)
San Francisco State University, San Francisco, CA, USA
e-mail: srosser@sfsu.edu

J. P. Carpenter (ed.), *Women in Academia*, Women in Engineering and Science,
https://doi.org/10.1007/978-3-031-97938-5_7

Although a growing literature, propelled partially by NSF programs such as ADVANCE (e.g., Gold et al., 2022; Rosser & Laursen, 2022; Zippel & Ferree, 2019), reports studies that have focused on gender in STEM faculty, most do not consider the FB status of the faculty, while those that study trajectories of FB faculty do not focus on gendered experience. This results in FB women STEM faculty being overlooked and understudied. While data on international students are publicly available (e.g., ASEE, 2023; NSB-NSF, 2021), collecting data for faculty becomes very difficult because of privacy issues and rapid changes in residency status; most university employers only track FB status of the faculty for visa purposes, without linking it to any internal assessments as they do with gender and URM status. A small number of researchers (e.g., Akulli, 2015; Foote et al., 2008; Rita & Karides, 2021; Wells, 2007) have studied the issues related to FB faculty in STEM.

Limited research has focused on the experiences of FB/FT faculty compared to the USB/UST faculty, and even fewer have used the intersectional lens of gender to explore issues faced by FB/FT faculty who identify as women compared to either FB/FT faculty who identify as men or USB/UST faculty who identify as women in STEM departments. While overall trends in recruiting and retaining more faculty who identify as women in STEM departments are promising, demographic markers such as URM status need to be carefully studied, especially with regard to FB status. Most of the foreign nationals are eventually naturalized and become citizens. While the immigration status of these faculty is transitional, their specific cultural and racial identify carries forward. Unfortunately, the classification of these individuals in URM/non-URM status is complicated (Rita & Karides, 2021) as 1) the URM definition used by NSF is based on underrepresentation in STEM fields relative to the overall US population, but FB faculty are drawn from the world population where the ethnic groups adversely affected by systemic inequities may or may not align with the US definitions; 2) FB faculty of Black and Hispanic backgrounds are included in URM, which raises the number of URM faculty but does not reflect an improvement in the including of historically underrepresented African American and Hispanic populations in the USA; and 3) FB faculty of White and Asian backgrounds are not included in URM because they are not underrepresented in STEM relative to their proportion in the the USA (although White and Asian women are underrepresented in some STEM fields), but many are still minorities in the education system and thus experience cultural isolation. Also, their experiences with bias are obscured when combined with USB non-URM experiences; Asian is an overly broad category that as a whole is overrepresented in STEM fields, but considering it as an aggregate masks pockets of underrepresented USB minorities (e.g., Hmong, Vietnamese), and White aggregates European, North African, and Middle Eastern, making it difficult to distinguish the biases they face. Furthermore, the definitions of whether Asians, who form the largest pool of PhD awardees in some areas such as engineering, are considered a racial minority vary between the NSF and U.S. Census Bureau, creating major gaps in understanding the demographic data on faculty in the professoriate.

7.2 Findings from a Systematic Review of Literature

To gain a better understanding of the experiences of FB/FT women faculty in STEM academic departments, during 2022–2023, I was privileged to earn an ARC Visiting Virtual Scholar (VVS) Award. As a VVS, I conducted a systematic review and meta-synthesis of over 3287 pieces of literature published between 2001 and 2021 on FB/FT faculty using intersectional lenses of gender and of race/ethnicity. The methodology and results of this research, including the rationale for the choice of metasynthesis, keywords used, inclusion/exclusion criteria, and databases searched, are published in a peer-reviewed article (Rosser, 2024) as well as in the report on the ARC's website (https://www.equityinstem.org/virtual-visiting-scholars#cohort5). The PRISMA diagram (Fig. 7.1) summarizes the sources of the 30 articles selected for final analysis:

The findings from the 30 articles, also informed by the 69 articles excluded primarily because they were quantitative, could be divided into six broad categories that capture facets of the academic experiences of FB/FT women faculty in engineering. Although the qualitative studies provide rich examples of issues facing FB women faculty, particularly women of color, which vary considerably, depending upon the country of origin of the individual and her immigration path to the USA,

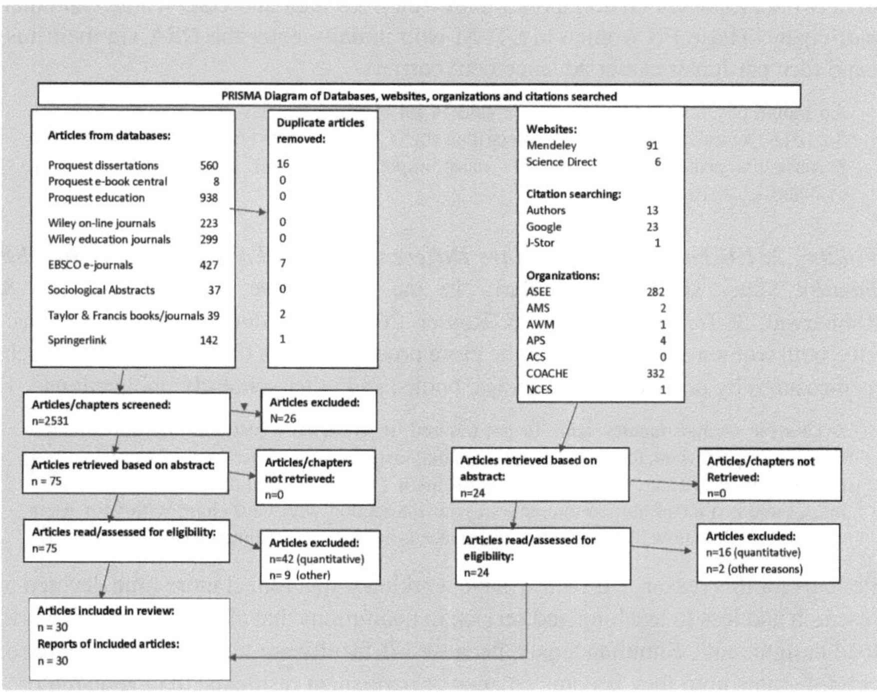

Fig. 7.1 PRISMA diagram of databases, websites, organizations, and citations searched (Source: Rosser, 2024)

six overarching findings emerge from the literature. Each of these findings presented below is described briefly with quotations from the qualitative literature to illustrate examples of each finding:

Finding 1: FB Women Appear to Have Different Experiences Compared to FB Men and to USB Women and Men Depending upon their race/ethnicity, FB women often experience the double bind of racism and sexism, which, when combined with discrimination that sometimes accompanies FB status, results in a triple bind.

> A Black African faculty member at a primarily white institution (PWI) stated: I encountered a lot of challenges when I first started. I spoke a slightly different accent coming from Africa! Students appeared disinterested when they started the class. Students later were amazed about where I gained knowledge because the image of Africa they have come to know is a poor struggling country with her nationals killing one another. I had to explain several times that Africa is a continent of 54 countries (Boafo-Arthur & Tsevi, 2021, p. 4).

Both USB faculty and students may have stereotypical expectations for women from certain cultures, for example, expecting women from some Asian cultures to be deferent and dress in particular ways. Some FB women have found that if FB men from their country of origin are also in their department or college, the FB men may treat them differently than USB women, based upon gender and/or religious expectations for women in their country of origin. FB women may not be included in networks for research collaboration in which FB men and USB women and men participate. Those FB women in STEM who initially enter the USA via their husband face particular career advancement barriers.

> An Indian physics associate professor said: A lot of women marry people senior to them. So, [if] he's more advanced in his career than she is, then she falls farther and farther behind because his professorship is always more important than her assistant professorship (Yakaboski, 2016, p. 76).

Finding 2: FB Faculty Tend to Have Different Workload Assignments than USB Faculty Many studies, particularly in the quantitative literature (Corley & Shbharwal, 2007; Mamiseishvili & Rosser, 2010), have documented that FB faculty, both women and men, tend to be more productive than USB faculty in research, as measured by peer-reviewed articles, books, and other scholarly publications.

> A Chinese woman faculty said: To get tenured, it is not only just getting your research funded and published, there are other things like visibility and leadership. All these kinds of things you need to start to consider; then I think it's all the cultural differences start to kick in … there is a lot of human interaction, your interaction with your chair, with your peers so that started to show like the culture difference is indeed there (Lightner-Noll, 2021, p. 80).

Perhaps for this reason, FB faculty have workloads that reflect more time devoted to research and less to teaching and service in institutions that allow differential workload assignments. Simultaneously, because FB faculty are not eligible to apply for certain grants until they become citizens or permanent residents, their research productivity may be inhibited.

> Denise, an assistant professor …: Being international has been a challenge because it prevents you from being able to apply for different grants. It is hard because you can have your good research to win that award but, without that paperwork, you can't apply for them. So, for my career, I have missed quite a few opportunities because of that status (Veliz 2014, p. 12).

Studies (Mamiseishvili, 2010, 2011; Mamiseishvili & Rosser, 2010) have documented that FB faculty produce fewer student credit hours; a few studies have suggested that the lower number of student credit hours results from lower numbers of students per class rather than assignments to teach fewer classes. Considerable research has documented that language issues and accents cause many students to avoid classes taught by FB faculty; this is particularly true for Asian-born faculty.

> A USB student said: "When scheduling classes, make sure your professor's name sounds American," I was told by several of my friends who were already in college. Several students reported having received similar advice "because school is already hard enough without having to struggle with trying to understand what your professor is saying" (Alberts, 2008, p. 192).

Many FB faculty discuss spending considerable time improving their English and preparing for classes, to overcome the language deficit they feel that they have.

Particularly if they did not receive any of their education in the USA, FB faculty struggle with understanding US higher education and its expectations for students.

> An African professor with experience in the European system summed these points up as follows: In Europe and Africa the only thing universities are interested in is to make people competent. If you are not competent, you don't graduate. Here you want to please the student, so it is a different scenario (Alberts, 2008, p. 197).

FB faculty tend to spend less time on service activities and to be assigned different service activities than USB faculty. In some cases, the service assignments FB faculty receive are focused on their FB status, for example, organizing events for international day or celebrations.

> Leela, a Full Professor from South Asia, discussed university efforts to create cultural awareness on campus. She said: This idea like multiculturalism, that's something I don't like. I don't like being invited as the [South Asian] in this multicultural event, sort of pigeonholed in a bracket where I am expected to get some [ethnic food] and wear a certain costume and show up. I don't like that because multiculturalism reinforces difference and makes it purely a matter of culture (Ghosh & Barber, 2021, p. 1071).

FB faculty are appointed to fewer "important" committees and those that are likely to result in leadership opportunities at the departmental, college, or university level. FB women faculty, especially those from Asia, feel that they are not considered for leadership positions.

> Almost at the end of her academic career, one foreign-born full professor realized the disadvantage of not being involved actively in administration: Now when I look back at my career, I could see a lack of leadership … And this leadership is difficult for a foreigner because I was raised in a traditional role, i.e., to be a little violet—you don't want to open your mouth with your accent. So, it's difficult. And many women don't know that they have to learn how to develop the skill how to run a meeting, how to be a leader (Skachkova, 2007, p. 717).

Finding 3: FB Faculty Tended to Cluster in R-1 Institutions Initially But Are Now Also in Comprehensive, Regional Universities Until the early 2000s, FB faculty were concentrated primarily in research-extensive and -intensive institutions, although some also found employment in liberal arts colleges. More recently, particularly with changes brought about by the Immigration Act of 1990 that expanded the number of immigrants, granted priority visas to highly skilled and educated workers, and created the temporary H-1B visas allowing employment, more international students received their PhDs from US institutions and remained in the USA for employment, particularly in STEM fields. Regional comprehensive institutions, 4-year private and public institutions, and even community colleges have now hired significant numbers of FB faculty.

> My university will not sponsor a green card till I get tenure; therefore, I am not eligible for early career funding from NSF or NASA compared to American colleagues. I had to take a job which emphasizes teaching (13–18 credit hours per semester) because I do not have the same access to research funds as my American colleagues (Theobold, 2009, p. 23).

Since China, India, and South Korea are the countries from which the largest number of international students in STEM come, that is reflected in the FB faculty recently hired.

Finding 4: US Immigration Policies, As Well As the Political Situation in the Country of Birth, Have Significant Impacts Upon Both the Numbers and Country of Origin of FB Faculty in the USA These policies and political changes create what economists define as pull-and-push situations that encourage increases in FB faculty in the USA. For example, in the wake of the fall of the Berlin Wall and the end of the Cold War, many FB faculty immigrated from Eastern Europe and former Soviet Bloc countries and became STEM faculty in the USA. The revolution in Iran pushed many Iranian scientists and scholars to immigrate to the USA and other countries.

> A faculty member from Iran stated: Things changed in Iran. We came here to get an education, and go back, and help with the revolution. But ... things changed. The revolution changed. Kids came along and ... this became our home basically. The last time that I went was 5 years ago, and each year I want to go but can't because of engagements here (Carattini, 2015, p. 5).

The U.S. Immigration Act of 1990 allowed many more international students to come to the USA on visas to study and then remain in the USA. In 2022, the Biden-Harris administration clarified that students on F-1 visas earning bachelor's, master's, and PhD degrees could remain in the USA for an additional 36 months to complete optional practical training if they were in "high-demand professional fields such as STEM." This has significantly increased the number of FB STEM faculty, especially from Asia. A few women discussed how difficulties of their visa processing affected their career decision-making.

> As an Indian national, Pema's visa retrogression triggered her job search though she was happy with her position at the time. She said: I think retrogression happened at that time. And the date, which was current, for the visa numbers, went back 5 years, which meant for people from India and China, so there was 10 Y. No way I could have been there for

> 4–5 years and so that was one of the reasons that triggered my job search again (Chena & Lawless, 2018, pp. 10–11).

Even FB faculty coming from European countries experience issues with immigration that can complicate their careers when transitioning to the USA:

> Ashley, a professor from a European country, shared the following: I was supposed to start my position in September and supposedly my visa was lost in the mail, so they had to ask for a copy. When it finally arrived somebody else was teaching my class … my teaching evaluations were horrible the first time I was teaching a class (Veliz 2014, p. 12).

Finding 5: Retention of FB Faculty, Especially Women, Depends Partly on Their Integration into the Broader Community As Well As Their Department and Institution Several studies have shown that FB faculty may experience difficulty with isolation at the university, although they may feel better integrated into their departmental, college, and university community compared to their integration into the surrounding community in which they live.

> Agnese, a woman faculty member from Southern Europe, for example, explained that men with whom she works "go out for coffee, or they have dinner parties at their houses, and they don't invite me." As the only woman in her department, she is isolated in ways that might appear informal and inconsequential but matter very much in terms of occupational mobility because these men will be the voting members on her tenure case (Ghosh & Barber, 2021, p. 1075).

This may be particularly true of FB faculty in small towns and rural communities where there may be very few FB individuals, especially professionals, not affiliated with the university. FB faculty who are perceived as racial/ethnic minorities in these settings may experience discrimination in the broader community, even when they feel accepted at the university.

> Also, that is one thing that helped me a lot adjusting to this culture and also we have faculty that are also like Asian and they really welcome us to some activities outside school, like inviting us to their parties and just getting us familiar with the environment. And gradually, once I passed those, I started to work with improving my language by talking to people, by talking to English native speakers (Lightner-Noll, 2021, p. 93).

FB women faculty in STEM may be especially isolated in the community, even if quite a few FB faculty from their country of origin are at the university because many of the FB women may not be professionals or even working because of visa issues if they entered the USA as wives of FB men faculty.

Finding 6: The Pull of Family and Friends in the Country of Origin Makes Many FB Faculty Feel That They Are Not At Home Either in the Country Where They Were Born or in the USA Absence of extended family and the expense and time required for international visits often make it difficult for FB faculty to see their family and friends in their country of origin as frequently as they would like.

> I think I put down more roots in the last 5 years than I had [earlier in my time in the United States] .... Until then, honestly, I would have had to say probably more than 50 percent of my heart was still with the family in India .... I [finally] found an emotional framework ...

> that I didn't have to such an extent [earlier in my time in the United States] (Carattini, 2015, p. 2).

FB women faculty from certain countries and cultures, where the expectation is that the women will foster close ties with the extended family, face a particular burden from such expectations.

> A woman faculty member from Asia explained: They do come visit us and stay for a couple months every year. My parents love it in the U.S. and so far it has been good but it is a concern in the future when they are older and you know, need more care. What I should do but that's something I haven't quite figured out yet (Lightner-Noll, 2021, p. 99).

Births, marriages, and deaths in the family, as well as holidays, exacerbate the feelings of estrangement from their extended family in the country of origin.

7.3 Discussion

These six findings characterize experiences expressed by many FB women faculty from a variety of countries and continents of origin. Some of the literature revealed more subtle findings that applied to particular racial/ethnic groups from certain countries. For example, several noted disparities in the experiences of White FB faculty compared to faculty of color (Ghosh 2015) from countries in the Caribbean (Hernandeza et al., 2015), Africa (Abla, 2012; Boafo-Arthur & Tsevi, 2021), and South Asia (Lawless & Chena, 2017) regarding discrimination and belonging. In some cases, tensions arose because of the institutional type, for example treatment of Black African vs. African American faculty at HBCUs (Afoaku, 2015; Ochukpue 2004). In other instances, the institutional setting in terms of region of the USA (Afoaku, 2015; Xue 2016), particularly in rural settings, contributed to increased discrimination for FB faculty of color.

Since very few of the studies focused on faculty from regional comprehensive (Roy, 2019; Oka et al., 2022; Lin et al., 2009; Muñoz, 2020) or 2-year institutions (Bista, 2016; Goodroad, 2019; Markus, 2011), it remains difficult to draw conclusions based upon Carnegie classification of institution, other than noting that FB faculty have increased dramatically beyond R-1 and R-2 institutions and liberal arts colleges where they predominated before the twenty-first century. One issue that contributes to this relative understudy of FB faculty outside of research institutions is that much of the qualitative data is found in PhD dissertations. Many of the students writing these dissertations study the populations present at the research institution where they are pursuing their degree.

Language, both written and spoken, becomes a significant issue for many FB faculty, particularly those from Asian countries. The written language can hinder research productivity, in terms of writing grants and publications; several faculty underlined the importance of collaborators or use of mentors (Canagarajah, 2018) that were native speakers to help with written English. Spoken English centers as a significant issue in teaching, often leading to poor student evaluations, complaints, and even students not signing up for courses if they know they will have a FB instructor (Alberts, 2008). Facility with spoken English may not suffice, as faculty from India and from England have noted student complaints in teaching evaluations about their accents. As indicated in one of the quotations about workload, concern over language and accent may also impact service due to inhibitions faculty impose upon themselves from speaking in meetings.

7.4 Gaps and Areas for Further Research

The results of the analysis confirmed the relative understudy of the significant portion of FB women faculty who make up the academic STEM faculty in US universities. In addition to neglecting this topic as an area for research in general, several aspects of the topic deserve particular focus and attention.

FB Faculty in Regional Comprehensive Universities and Community Colleges As indicated earlier in this chapter, most of the research on FB faculty has focused on faculty in research institutions. Although initially and especially before the twenty-first century, most of the FB faculty were concentrated in research institutions, this is no longer the case. Increasingly, FB faculty now have obtained tenure-track positions in regional comprehensive institutions and community colleges; in some departments and disciplines such as engineering and computer science, they constitute the predominant percentage of faculty in such institutions. Yet very few studies have focused directly on FB faculty in such institutions. While faculty in such institutions may be included in some studies, typically quantitative studies using large samples or databases, often the data are not disaggregated in an intersectional way that yields results specifically for FB women faculty, especially women of color, in such institutions.

This focus on FB faculty at research institutions opens the possibility that some of the major findings surrounding FB faculty may not hold in environments outside of research institutions. For example, the finding that the FB faculty tend to have higher research productivity may not be the case in institutions where the major criterion for initial hiring, as well as promotion and tenure, is not research. Similarly, the findings surrounding FB faculty production of fewer student credit hours may not occur in regional comprehensive and/or community colleges where hiring and advancement of faculty are centered on teaching expertise, rather than research. Faculty in regional comprehensive and community colleges often also are unionized, depending upon the state and other contextual parameters; union contracts may constrict or impact in major ways the leeway permitted for variations in faculty workload assignment. More research on faculty in such institutions would clarify the issues for FB faculty in those environments and the extent to which they overlap or diverge from those of faculty in research institutions.

Impact of US Policies on Future Mixture of Faculty Currently Predominant in STEM Departments Federal policies need to anticipate the future impacts of the China initiative and other US policies and political action on STEM, since the largest number and percentage of FB graduate students and FB/FT faculty currently come from China.

Impact of COVID on Careers of FB Women Faculty Although some of the research coming out about the impact of COVID on women faculty in STEM indicates more research disruptions and fewer publications than those experienced by their male colleagues, to my knowledge, no research has focused on the impact of COVID on FB women faculty compared to their USB women or men colleagues or FB male colleagues. In addition, it could be important to find out whether COVID made FB women faculty less likely to continue their careers in the USA or return to their home country of origin.

Career Trajectories of FB Women Faculty Much of the limited research on FB women faculty has focused on entry and mid-career FB women faculty. More research is needed to understand the experiences of FB women faculty and the mechanisms that enable their success as they mature in their careers and progress though the life span. These longitudinal studies will be especially salient as the influx of FB women in STEM from Asian countries increased dramatically when the immigration laws changed to allow more students from these countries to pursue graduate studies and remain in the USA to pursue employment.

Tension Between FB Women Faculty and FB Men Faculty from Same Country of Origin in the Same Department Anecdotal information and rumors suggest that FB women faculty occasionally face discrimination based upon stereotypical expectations regarding gender roles from FB men faculty from their same country of origin. Although these FB men faculty may have learned not to place these expectations on USB women colleagues, they may continue to hold and enforce these gender-role expectations for their FB women colleagues. Research needs to be undertaken to understand whether these rumors and anecdotes are based on facts, since virtually no studies have explored this dimension of FB women faculty's experience.

Hiring of FB Couples in STEM Departments Although one study (Yakaboski, 2016) focuses explicitly on the hiring of FB couples in STEM departments and a few other works allude to this as an issue in retaining FB women, little research has focused specifically on this topic. Since this is a well-known issue in the literature on USB women in engineering, this needs to be explored to understand better the differences between FB and USB faculty on this topic.

7.5 Policy Implications for Higher Education Institutions

The literature points to a number of policy and practice implications to improve the experiences of FB/FT faculty and enhance success and retention. Some of the policies are relevant for federal agencies funding STEM, such as the suggestion that NSF overhaul its definition of URM, since not all Asian countries and Asian women are underrepresented in some STEM disciplines and since the current definition may lead to the perception that USB African American and Hispanic faculty are increasing due to Black and Hispanic FB faculty being classified as URMs after they obtain permanent residency or US citizenship. Others are relevant for current policies and practices of the federal government that are likely to influence the number of international students and faculty from certain countries such as China and likely to impact future STEM enrollments and workforce (Quinn, 2023).

In this chapter, I focus on implications for policies pertaining to higher education at the departmental, college, and institutional levels:

- *Orientation:* Issues and challenges that FB faculty may face should be added as an aspect of new faculty orientation received by all new faculty, both FB and USB. This will not only facilitate the transition of FB faculty to institutions of higher education but will also provide USB faculty some insights into particular issues faced by their FB colleagues.
- *Mentors:* Each FB faculty member should have at least one mentor, including an individual who was FB and transitioned to become a faculty member in higher education in the USA. A repeated recommendation that emerged from the literature (Bloom, 2017; Phillips et al., 2016) was the importance of mentoring. Although mentoring is important for all new faculty, it becomes especially significant for FB faculty adapting to a new country and educational system. Ideally, a FB faculty would have more than one mentor, some who are USB and at least one that is FB.
- *Immigration issues:* Help and guidance needs to be provided with immigration issues. In addition to a mentor to help guide the faculty member with immigration issues, having a university office that includes individual(s) with knowledge and expertise in immigration issues serves as a critical aid for this process. All deans and department chairs should also have some knowledge of immigration issues, at the very least knowing the individuals on campus with expertise to help, and understanding the impact that immigration issues have on the FB faculty member's progress towards tenure and promotion.
- *Social isolation:* Deans and chairs should be cognizant that FB faculty face the potential of social isolation and that they need to provide opportunities to increase their social interaction with colleagues. This becomes particularly crucial for women faculty who may be the only woman in an engineering department and may be excluded from informal networking and collaboration opportunities. Faculty on rural campuses located outside of large urban areas may face isolation from social opportunities and active discrimination in the broader community, even if they feel included in the university.

7.6 Conclusion

Although the number and percentage of FB/FT faculty have increased dramatically in academic departments throughout the USA during the last two decades, so that they constitute a majority of faculty in many engineering departments, very little research has focused on the experiences of these faculty, particularly the FB/FT women. This systematic review reveals that FB women appear to have different experiences and different workload assignments and that their retention depends partly on their integration into the broader community as well as their department and institution. US immigration policies, along with the political situation in the country of birth, have significant impacts upon both the numbers and country of origin of FB faculty in the USA, who tended initially to cluster in R-1 institutions. The pull of family and friends in the country of origin makes many FB faculty feel that they are not at home either in the USA or their country of birth. These findings might be used to determine policies and practices to facilitate FB/FT women's success in academic workplace environments evolved and traditionally populated by USB faculty. The knowledge resulting from these findings could be used to encourage FB/FT and USB/UST faculty to realize differing perceptions of each other's experiences in the workplace, stimulate scholars to research the gaps in knowledge, and allow administrators to better address shortcomings in the work environment.

References

Abla, Z. W. (2012). Portraits of successful African immigrant faculty on US campuses. Ph.D. Dissertation. Bowling Green State University.

Afoaku, O. H. (2015). The experiences of African-born women faculty and administrators at colleges and universities in the United States. Ph.D. Dissertation, Indiana State University.

Akulli, A. N. (2015). Embeddedness of foreign-born faculty at comprehensive universities in the United States. Ph.D. Dissertation, Michigan State University.

Alberts, H. C. (2008). The challenges and opportunities of foreign-born instructors in the classroom. *Journal of Geography in Higher Education, 32*(2), 189–203. https://doi.org/10.1080/03098260701731306

American Society for Engineering Education (ASEE). (2023). *Profiles of Engineering and Engineering Technology, 2022*. Washington, DC. ARC website. Retrieved from https://www.equityinstem.org/virtual-visiting-scholars#cohort5

Bista, K. (2016). In R. I. Raby & E. J. Valeau (Eds.), *Faculty International Experience and Internationalization Efforts at Community Colleges in the United States*. International Education at Community Colleges.

Bloom, L. (2017). New faculty mentoring in higher education: An integrated literature review. Ph.D. Dissertation, Concordia University—Chicago.

Boafo-Arthur, S., & Tsevi, L. (2021). Experiences of Black African international faculty at American higher education institutions. In C. Glass, K. Bista, & X. Lin (Eds.), *The Experiences of International Faculty in Institutions of Higher Education: Enhancing Recruitment, Retention and Integration of International Talent*. Routledge.

Canagarajah, S. (2018). English as a spatial resource and the claimed competence of Chinese STEM professionals. *World Englishes, 37*, 34–50.

Carattini, A. (2015). Career trajectories and incorporation strategies in the life histories of foreign-born faulty in the U.S. Ph.D. Dissertation, University of Maryland.
Chena, Y., & Lawless, B. (2018). "Oh my god! You have become so Americanized": Paradoxes of adaptation and strategic ambiguity among female immigrant faculty. *Journal of International and Intercultural Communication, 11*(1), 1–20.
Corley, E. A., & Shbharwal, M. (2007). Foreign-born academic scientists and engineers: Producing more and getting less than their U.S.-born peers? *Research in Higher Education, 48*(8), 909–940. https://doi.org/10.1007/s11162-007-9055-6
Foote, K. E., Li, W., Monk, J., & Theobald, R. (2008). Foreign-born scholars in US universities: Issues, concerns, and strategies. *Journal of Geography in Higher Education, 32*(2), 167–178. https://doi.org/10.1080/03098260701731322
Ghosh, D., & Barber, K. (2021). The gender of multiculturalism: Cultural tokenism and the institutional isolation of immigrant women faculty. *Sociological Perspectives, 64*(6), 1063–1080.
Ghosh, D. (2015). US higher education in the construction of "good" and "bad" immigrants: Foreign-born faculty's tales of sexism, racism, and xenophobia at work. Doctoral Dissertation, Sourthern Illinois University at Carbondale.
Gold, J., Gates, A., Haqu, S., Melson, M. C., Nelson, L. K., & Zippel, K. (2022). The NSF ADVANCE network of organizations. *ADVANCE Journal, 3*(1). https://doi.org/10.5399/osu/ADVJRNL.3.1.3
Goodroad, E. (2019). Experiences of foreign-born faculty in U.S. two-year colleges. Ph.D. Dissertation. Edgewood College.
Hernandeza, K. C., Ngunjirib, F. W., & Changa, H. (2015). Exploiting the margins in higher education: A collaborative autoethnography of three foreign-born female faculty of color. *International Journal of Qualitative Studies in Education, 28*(5), 533–551. https://doi.org/10.1080/09518398.2014.933910
Lawless, B., & Chena, Y. (2017). Multicultural neoliberalism and academic labor: Experiences of female immigrant faculty in the U.S. academy. *Cultural Studies ↔ Critical Methodologies, 17*(3), 236–243. https://doi.org/10.1177/1532708616672688
Lightner-Noll, J. (2021). Cultural adjustment and career navigation of international faculty women in STEM: U.S. transitions and Confucianist influences. Ph.D. Dissertation, The Kent State.
Lin, Z., Pearce, R., & Wang, W. (2009). Imported talents: Demographic characteristics, achievement and job satisfaction of foreign born full time faculty in four-year American colleges. *Higher Education, 57*, 703–721. https://doi.org/10.1007/s10734-008-9171-z
Mamiseishvili, K. (2010). Foreign-born women faculty work roles and productivity at research universities in the United States. *Higher Education, 60*, 139–156. https://doi.org/10.1007/s10734-009-9291-
Mamiseishvili, K. (2011). Teaching workload and satisfaction of foreign-born and U.S.-born faculty at four-year postsecondary institutions in the United States. *Journal of Diversity in Higher Education, 4*(3), 163–174.
Mamiseishvili, K., & Rosser, V. J. (2010). International and citizen faculty in the United States: An examination of their productivity at research universities. *Research in Higher Education, 51*, 88–107. https://doi.org/10.1007/s11162-009-9145-8
Markus, L.C. (2011). Job Satisfaction of Foreign-born Faculty in Community Colleges Using NSOPF 2004 Data. Ph.D. Dissertation. North Carolina State University.
Muñoz, E. E. (2020). Attitude is (mostly) everything: A multi-institutional exploration of organizational climate. Thesis for Master of Science in Biology, California State University, Fresno.
NSB-NSF. (2021). The STEM labor force of today: Scientists, engineers and skilled technical workers. In *Science and Engineering Indicators 2022*. National Science Board, National Science Foundation. NSB-2021-2. Retrieved from https://ncses.nsf.gov/pubs/nsb20212
Ochukpue, W.E.N. (2004). A qualitative study of African immigrant professors in two historically Black institutions in a Southeastern state. Doctoral Dissertation, North Carolina State University.

Oka, L. G., Rosser, S., Mazari, M., & Abercromby, K. (2022). Investigating tenure experiences of foreign-born women faculty in engineering at the CSU system. In *ASEE Annual Conference, Minneapolis, MN, June 26-29*. Paper ID #37802.

Phillips, S. L., Dennison, S. T., & Davenport, M. A. (2016). High retention of minority and international faculty through a formal mentoring program. *To Improve the Academy, 35*(1), 153–179.

Quinn, R. (July 5, 2023). *Chinese scientists increasingly leaving U.S.* IHE.

Rita, N. P., & Karides, M. (2021). 'I have an accent, so people know I'm not from here': A racial and ethnic analysis of international STEM faculty in Hawai'i. *Ethnic and Racial Studies, 45*, 1–23. https://doi.org/10.1080/01419870.2021.1981965

Rosser, S. V. (2024). Towards a better understanding of the experiences of FB/FT STEM faculty: A systematic review and qualitative metasynthesis of the literature through intersectional lenses of gender and race/ethnicity. *ADVANCE Journal.*

Rosser, S. V., & Laursen, S. (2022). Looking back to look forward: A retrospective examination of ADVANCE. *ADVANCE Journal, 3*(1). https://doi.org/10.5399/osu/ADVJRNL.3.1.2

Roy, S. (2019). Experiences of Asian American and immigrant Asian women in the United States higher education administration. Ph.D. Dissertation, University of North Florida. Retrieved from n00961741@unf.edu

Skachkova, P. (2007). Academic careers of immigrant women professors in the U.S. *Higher Education, 53*, 697–738. https://doi.org/10.1007/s10734-005-1976-4

Theobold, R. (2009). New faces in academic places: Gender and the experiences of early career foreign-born and native-born geographers in the United States. *Journal of Geographical Sciences, 57*, 7–32.

Veliz Calderon, D.X. (2014). The socialization of international women faculty. Ph.D. Dissertation, The University of Maine.

Wells, R. (2007). Summer © 2007 Wiley Periodicals, Published online in Wiley InterScience. Retrieved from www.interscience.wiley.com). International faculty in U.S. Community Colleges. *New Directions for Community Colleges, 138*, 77–82. https://doi.org/10.1002/cc.284

Xue, M. (2016). Professional career navigation of East Asian international faculty: Constraints, enablements, and agency. Doctoral Dissertation, The University of Alabama.

Yakaboski, T. (2016). Reframing the two-body problem in U.S. STEM departments: Asian women faculty negotiation of career and family. *Journal of the Professoriate, 8*(2).

Zippel, K., & Ferree, M. (2019). Organizational intersections and the creation of gendered knowledge: U.S. universities and NSF ADVANCE. *Gender, Work and Organization, 26*(6), 805821.

Chapter 8
Women Leading Engineering and Computing Professional Societies

Karan L. Watson and Stephanie G. Adams

8.1 Introduction

The diversity of structure, clientele, membership, and purpose of professional societies related to engineering has significant variations. Some focus on personal demographic identities, some on fields of engineering, some on particular functions like education or accreditation, some on regional location, and more. In addition, some have long histories, some have huge staffs or no staff, and some have a number of funding resources, such as individual dues, institutional dues, services sold, publications sold, and grant funding. But what all have is their own unique organizational culture, which involves not only artifacts like those listed above, but also espoused values and missions and underlying assumptions about operations and leadership. Before seeking a leadership role in any of these societies or associations, an aspiring leader must strive to understand the organization's culture and how their own personal identity can align with that culture. We will give a brief perspective on women in leadership and on a general model of professional societies. These factors are essential to understand the challenges for women ascending to leadership roles in professional societies.

K. L. Watson (✉)
Texas A&M University, College Station, TX, USA

S. G. Adams
University of Texas at Dallas, Richardson, TX, USA

J. P. Carpenter (ed.), *Women in Academia*, Women in Engineering and Science,
https://doi.org/10.1007/978-3-031-97938-5_8

8.2 Women in Leadership

When more women are empowered to lead, everyone benefits. Decades of studies show women leaders help increase productivity, enhance collaboration, inspire organizational dedication, and improve fairness. Despite these benefits, only 10% of Fortune 500 companies are led by women (Novotney, 2023). According to data from the American College President Study (Amer. College Pres., 2023), produced by the American Council on Education (ACE), men outnumbered women two-to-one in the presidency. The representation of women presidents has increased from 23% in 2006 to 32.8% in 2022 (Melidona et al., 2023). Five major reasons for this low representation include gender bias, lack of role models, work-life balance, stereotypes and expectations, and limited access to networks and opportunities (Semeshkina, 2023). These factors play a role in women's presence in significant leadership roles in business and industry, education organizations at all levels, and nonprofits (Leanin.Org and McKinsey & Co., 2022).

In Fig. 8.1, the distribution of women in leadership positions for university positions shows that women often hold support positions that may not have authority over faculty lines. Similarly in engineering programs in higher education in the USA, the highest percentages of women in leadership roles were assistant deans (i.e., 43%) and associate deans (i.e., 31%). Both positions have higher percentages of women working in those roles than the percentage of women in engineering faculty positions. However, only 13% of department heads/chairs within engineering programs were women (Roy, 2021). These data on women in academia are shown at a time when among tenured/tenure-track faculty for 2022, women comprised

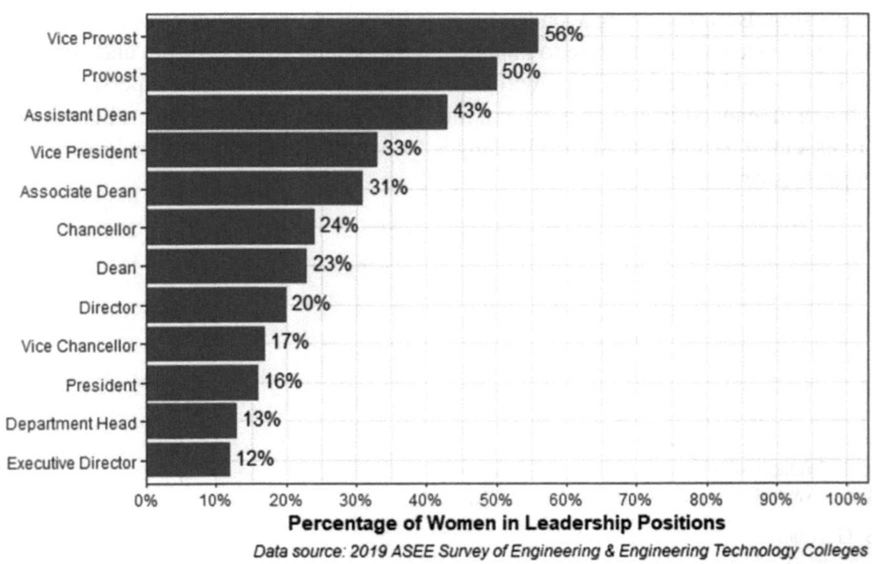

Fig. 8.1 Percentage of women in leadership positions (Data source: 2019 ASEE Survey of Engineering & Engineering Technology Colleges)

24.5% of assistant professors, 20.8% of associate professors, and 13.1% of full professors.

With the many barriers to women in leadership roles, it may seem counterintuitive that the opportunity to lead a national engineering society would be a strategic leadership goal for women. However, we posit that this climb in such an organization offers valuable learning opportunities that may be more difficult but more enlightening for a leader, of any gender, but especially for women. The five major barriers for women mentioned by Semeshkina are amplified in different ways in these leadership roles and can enlighten the woman to these factors and how to manage them in their employment opportunities.

8.3 Gender Vector Model for Engineering Professional Societies

Professional societies related to engineering tend to be oriented toward a particular field of engineering (e.g., ASME, IIE, IEEE), functions in engineering (e.g., ASEE, WEPAN, ABET, NSPE, Society of Fire Protection Engineers, Vertical Flight Society), honorific organizations (e.g., Tau Beta Pi, NAE, Eta Kappa Nu), or personal identity (e.g., SWE, NSBE, SHPE, AISES). Other orientations include regional, national, or international; new or well established; focused agenda; or broad inclusivity. However, within all these organizations are some artifacts that women must be particularly tuned to, which include the gender orientation (specifically what percentage of leadership is male) of the society's membership, the percentage of memberships that are male, and the gender diversity of the organizations who employee the society's members.

In the graph shown in Fig. 8.2, on any axis, the more male-oriented the composition, the greater the axis value. Thus, when a society is roughly plotted in this space, the greater the male orientations, the longer the vector for a woman to encounter in

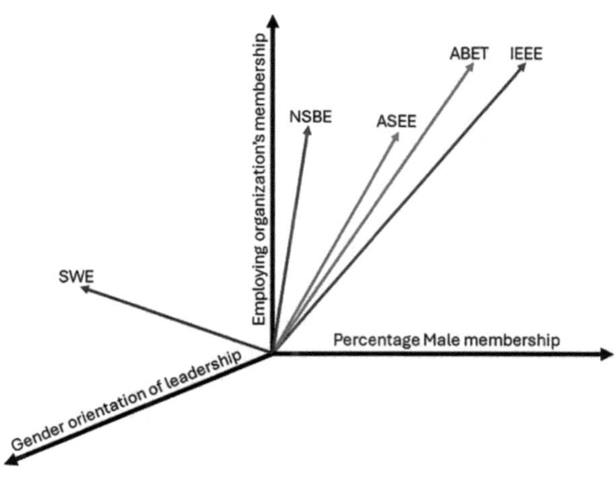

Fig. 8.2 Gender vector model for engineering Professional Societies

a leadership role due to their identity as a woman. This does not automatically mean that the leadership role poses a greater challenge for a leader who may be dealing with challenges due to issues such as staffing competencies, financial challenges, membership challenges, interorganizational political upheavals, etc. However, the woman leader is going to need to calibrate their personal sensitivities to the overall gender orientation of the society in order to be an effective leader. The figure also shows the gender vector for the examples described below (note that the examples use data gleaned from organizations' websites and represent data that is available from the last 5 years).

The Institute of Electrical and Electronics Engineers (IEEE) and its predecessors date to 1884 and have over 420,000 dues paying members from over 190 countries (64% outside the USA), and the primary source of funding is its technical publications and events. It remains focused on the clientele who are from fields that tend to be more than 80% male, has a membership that is estimated to be approximately 12% women (women are 30% of student members, 9% members, 8% senior members, and 4.4% fellows), and has only had four women presidents, all in the last 30 years. It has its first ever executive director who is female, after she served several years as IEEE's CFO. IEEE has a very long vector for women in leadership in our organizational space.

American Society for Engineering Education (ASEE) and its predecessor were founded in 1893; it is focused primarily on the teaching of engineering, where women comprise approximately 17.5% of teaching faculty. It has had ten women presidents in the last 30 years, but none apparently before that; currently, it has a woman executive director and has over 12,000 members. The current membership breakdown is 47.9% male, 23.27% female, 28.2% providing no response, 0.3% non-binary, and 0.3% preferring not to disclose (ASEE, 2023). ASEE has a medium vector length in our organizational space.

Society of Women Engineers (SWE) is over 70 years old, has significant representation of women in staff leadership, and has over 40,000 members, who are 90% USA located, 40% from academia, 3.3% male, and from many engineering disciplines. The presidents of SWE have overwhelmingly been women. SWE has a short vector in our organizational space.

ABET (formerly Accreditation Board of Engineering and Technology) and predecessor are approximately 90 years old; it does not have individual members but has 35 member societies who provide over 2200 volunteers to accredit 4564 programs at 895 colleges and universities in 40 countries. ABET has had five women presidents in the last 30 years, and none before that, and has always had a male-dominated leadership staff. ABET has a high vector in our organizational space.

National Society of Black Engineers (NSBE), the largest student-governed organization based in the United States, turned 50 in 2024 and strives *to increase the number of culturally responsible Black engineers who excel academically, succeed professionally, and positively impact the community* (NSBE, 2023). Comprised of students (K-12, undergraduate, and graduate) and alumni professionals, NSBE has over 23,000 members with 39.22% identifying as women. NSBE has a medium/high vector in our organizational space.

The point of the vector length is to give a woman leader some idea of the organizational culture in which they will operate as a leader. There are many aspects of organizational culture that any leader, whatever their gender, must deal with, but here we want to focus on the particular challenges women leaders may face.

8.4 Considering a Society's Gender Vector When Leading

Any organization's leader was probably selected because enough people involved in the selection believed that leader "fit" with the current needs of the culture of the organization. This is true whether they wanted to see the maintenance of the culture or a change in the culture. Although people normally involved in the selection of leaders have a singular perspective of the organization, the process may involve people with varied perspectives. The selected leader also has a singular perspective of the organization, but it behooves them to reflect on the multiple perspectives.

A framework for doing this is Edgar Schein's model of organizational cultures (Schein, 2010). This model makes you to consider the organization's visible artifacts (i.e., composition, processes, mission, products, and honors) as you consider the organization's espoused values (i.e., vision statements, diversity statements, core values). In considering these aspects of the organization, and how well artifacts align with espoused values, a person can start to understand the shared underlying beliefs and assumptions that really motivate how the organization operates and who thrives in the organization. The gender vector model above highlights artifacts of an organization that begins to uncover important underlying assumptions about the organization.

A woman in leadership not only must understand the same aspects of an organization's culture as a man does but also needs to be keenly aware of the organization's macro- and micro-assumptions about women's roles and expected behaviors. She must not presume that all organizations' cultural norms have the same underlying assumptions about women in leadership. A woman may be sensitized to such awareness, whether consciously or unconsciously by her work culture, and now is being asked to lead in a different organizational culture (the society). For example, many within the student affairs context of higher education are comfortable sharing pronouns (they/them, he/him/his, or she/her/hers), while many in nonstudent affairs units are hesitant, reticent, or unwilling. Women leaders must not assume that all organizations have the same underlying assumptions about women in leadership.

All organizations may have sexist, racist, or other negatively biased members. The organizational culture determines how they survive and lead (overtly or covertly) with such prejudices. In addition, all organizations have to deal with implicit and unconscious biases. The organizational culture determines whether members are aware and practiced at overcoming these biases or unaware and unchecked on such biases. The woman leader is not being asked to not be herself, but for success as a leader must understand the "waters" in which she is swimming in. Or, in the language of the "Art of War for Women," she must understand the "ground" on which she must take her stand (Chu, 2010).

Understanding the gender vector of the society in which she will lead and contrasting this vector with a similar vector of other organizations and how leaders operate and thrive in those organizations will inform the leader significantly. She may need to curb her reactions and confrontations with biases encountered in the society. She may need to have allies take on some needed messages or changes for the society, while she addresses more pressing challenges or visions for her leadership. To illustrate these ideas, we describe personal cases.

8.5 Cases Taking Gender Vector into Account

One author was elected as the President of ABET, as mentioned above, an organization with a high gender vector for women. At the time of her election, she had served at leadership levels in a college of engineering (high gender vector) and university leadership levels of an organization with a medium gender vector. Thus, her work experiences had prepared her for an organization like ABET. Furthermore, she had served as a program evaluator for IEEE (very high gender vector) and a member of the engineering accreditation council (high gender vector) and on the ABET board (high gender vector). Thus, she was well prepared from a gender perspective for the leadership of ABET.

In her term, the ABET board had been discussing the functional effectiveness of the board because of its size (65 members) and a persecution that board members felt they represented their nominating society more than the ABET organization. Some recent factors also included that the board had discussed these concerns several times previously and failed to make changes, and instead the board had been enlarged recently by providing nonvoting membership for some professional organizations that represented women or underrepresented minority engineers.

The challenge of getting board members to vote themselves out of seats on the board was obviously huge, and probably not a 1-year task (the term length of ABET presidency). In addition, some of the organizations most opposed to any change came from societies with very high gender vectors and felt that it was hypocritical to talk about a board being too big when expansion for women and minority representing organizations had recently been added. The author was known to have voted for these additions.

With her experiences within her work organization and prior service in ABET, some of her strategies to bring about board changes were the following:

- Knowing board members often sidetracked conversations with Robert's Rules of Order and ridiculed people who were not adhering to those rules, and this would be used in harsh ways against a woman (the author, not an expert in Robert's Rules) who got the board to adopt simplified rules for these discussions and had two volunteers, both males and one for the change discussion and one against, who would monitor the process of discussion.

- The author and the president-elect, a male, announced that they were both in agreement that this discussion needed to happen and that we would take 2–3 years to discuss before a decision would be called for. The president-elect agreed that even during his term, the author would lead these discussions, and the following president-elect also agreed that the author would lead the discussion if it took until her term as well.
- In organizing the discussions among board members, the author paid attention to forming discussion clusters that were diverse along many aspects of the organizations' diversity (not simply their percentage of membership that is female). These aspects included such things as size of membership, number of programs they represented, diversity of their board members, as well as their gender vector, and those leaning in favor and against the changes under consideration.
- The author asked board members, male and female, who had been for the addition of the identity-based organizations on the board to ensure that their voices were valued in the discussions so that the author did not have to be the main advocate for their voices.

Any leader may have taken these same steps; however, as a woman leader, the first in 20 years, the author knew that these steps would help lessen the high gender vector (in relation to her leadership) in having undue influence on the discussion.

The second author was elected President of ASEE, a medium gender vector organization. Her narrative on that journey:

She attended her first ASEE conference in the summer of 1996 as a graduate student, and Eleanor Baum, Dean of Engineering at Cooper Union, was the President and the first woman to hold that position after 102 years of ASEE's existence. It would be another 9 years before the second woman Sherra E. Kerns would become President. As previously mentioned, a total of ten women have held the title of ASEE President over the past 30 years, including six women serving consecutively as President from 2017 to 2023. In 2022, ASEE appointed a woman as Chief Executive Officer, the first in the history of the organization.

When she joined ASEE as a faculty member, she became an active member of the engineering management division and would go on to hold leadership positions including immediate past chair, division chair, program chairperson, treasurer, and secretary, from 2001 to 2006. In 2010, she was elected Professional Interest Council I (PIC) Chair, a role she held for 3 years from 2010 to 2013. During 2011–2012, she also held the position of Vice President, PICs. In 2013, she was elected as a Fellow of the organization. Between 2013 and 2018, she served the Society on several ad hoc committees, working groups and task forces, and standing committees. In 2018, she was chosen as President-Elect by her peers, which led to her service as President from 2019 to 2020. She was the seventh woman to hold this position and the second Black/African American woman to do so. When she joined the Board in 2010, it was comprised of nine women and ten men. When she served as President, the board had ten women and nine men.

It is unclear what challenges President Baum might have faced as the first woman president of ASEE, but this author can say with certainty that she does not believe she faced any additional scrutiny or challenges as a board member or as president. Throughout her involvement and service over the past 28 years, I found a welcoming community with a passion for teaching and for creating new experiences for our students and colleagues, a space to try and report new pedagogies or experiences in research and teaching, and an overall nurturing environment that allowed her to refine her thoughts on leadership and service. She saw women in different levels of leadership throughout her journey, both as a board member and among the staff. Both continue to maintain decent gender representation. Based on personal experience, this author confirms the medium vector assignment for ASEE.

The second author has also had a 40-plus-year relationship with the National Society of Black Engineers. When she entered college in 1983 at the University of Wisconsin-Madison, she was introduced to the Wisconsin Black Engineering Student Society (Wisconsin's chapter of the National Society of Black Engineers). Founded at Purdue University in 1975 by six male students, Edward Coleman, Anthony Harris, Brian Harris, Stanley L. Kirtley, John W. Logan Jr., and George Smith, NSBE elected Virginia Booth as its first female national chair (NSBE's term for president) in 1978. Booth would go on to become the first chair to serve two terms, and she currently serves on the National Advisory Board. Throughout NSBE's 50-year history, 43 individuals have been elected chairperson, of which 11 or 26% were women.

In 1987, Rev. Dr. Florida Morehead (Executive on Loan) became the first full-time Executive Director. In 1989, Morehead would become an NSBE full-time employee. Since 1987, there have been a total of six executive directors with 50% being women.

In 1989, the author was elected to the National Executive Board where she served 3 years (two as Publications Chair and one as Vice Chairperson). She would hold other board positions or serve on the National Advisory Board for over two decades. Once again, this author does not feel as though she experienced any difficulties based on her race or gender.

Perspectives shaped by intersectional and lived experience enhance decision-making processes and develop a culture of inclusivity within organizations. According to the 2023 Women in the Workplace report, women who demonstrate the ability to navigate challenges and excel in leadership positions create an environment where all individuals, regardless of their background, are motivated to achieve their best.

The challenges faced by Black women on the path to leadership are substantial and can seriously hinder their progression. As Rachel states, Black women deserve to be acknowledged for the value they bring. It is essential to recognize our track record and accomplishments, ensuring that our contributions play a vital role in the success of an organization. Being heard and having our opinions valued are an integral part of this acknowledgment (Leanin, et al. 2022).

8.6 Conclusion

It has been shown time and time again that women leaders create environments that acknowledge diverse perspectives, advocate for initiatives and policies that advance equal opportunity and inclusivity for all constituents, and foster a more inclusive and supportive culture. Women leaders have also been shown to tackle intricate challenges while also fostering robust interpersonal connections and facilitating the efficient exchange of ideas.

The presence of women on leadership teams contributes to a wider array of experiences and perspectives, resulting in decision-making that is more informed, innovative, and efficacious. Due to this diversity, professional societies can adapt to shifting environments and make more informed decisions that benefit their members and the community.

As former National Academy of Engineering President Bill Wulf said, "Every time an engineering problem is approached with a pale, male design team, it may be difficult to find the best solution, understand the design options, or know how to evaluate the constraints (Wulf, 1998)." Gender-diverse leadership facilitates the development of innovative and efficient problem-solving strategies. When leadership is nonhomogeneous, challenges are approached from various perspectives, resulting in more comprehensive solutions that may not be readily apparent. The presence of diverse leadership within the professional society guarantees that a wider array of members' needs and concerns are taken into account and resolved, thereby fostering the development of policies, programs, and initiatives that are more inclusive.

References

American Society for Engineering Education (ASEE). (2023). Profiles of engineering and engineering technology, 2022.

Chu, C. (2010). *The art of war for women: Sun tzu's ultimate guide to winning without confrontation*. Crown Publishing Group.

Leanin.Org and McKinsey & Company. (2022). Women in the Workplace 2022. Retrieved from https://www.mckinsey.com/~/media/mckinsey/featured%20insights/diversity%20and%20inclusion/women%20in%20the%20workplace%202022/women-in-the-workplace-2022.pdf

Melidona, D., Cecil, B. G., Cassell, A., & Chessman, H. M. (2023). American College Presidents Study.

National Society of Black Engineers. (2023). Retrieved from https://nsbe.org. Accessed 22 May 2023

Novotney, A. (2023). Here's the science behind how to promote them Psychological research shows women leaders improve businesses. Experts share how to increase the number of women in leadership roles. Retrieved from https://www.apa.org/

Roy, J. (2021). *Women in Leadership Positions*. ASEE. Retrieved from https://ira.asee.org/women-in-leadership-positions/#:~:text=Within%20engineering%20programs%2C%20the%20highest,women%20in%20engineering%20faculty%20positions

Schein, E. H. (2010). *Organizational Culture and Leadership* (4th ed.). John Wiley and Sons.
Semeshkina, M. S. (2023). Five reasons why we don't see more women in leadership roles. Retrieved from https://www.forbes.com/sites/forbesbusinesscouncil/2023/03/13/five-reasons-why-we-dont-see-more-women-in-leadership-roles/?sh=28352e38574b
The American College President. (2023). *Edition—Executive Summary*. American Council on Education.
Wulf, W. A. (1998). The urgency of engineering education reform. *Bridge, 28*(1), 48.

Chapter 9
Academic Women in the National Academy of Engineering

Jenna P. Carpenter

9.1 Early History

The first (academic) woman admitted to the National Academy of Engineering (NAE) was Lillian Gilbreth in 1965, just one year after its founding (National Institutes of Health, n.d.). Her election citation reads "*motion study and management engineer.*" She was a part-time Professor of Management at Purdue University from 1935 to 1948 (National Academy of Engineering, n.d.-a). The Society of Women Engineers has a list of all women elected to the NAE by year and by discipline, Figures 9.1 and 9.2 (Society of Women Engineers, 2024a), as well as by name (Society of Women Engineers, 2024b), all current through 2024.

A search of all of the National Academies of Sciences, Engineering, and Medicine (NASEM) committees since 1950 with "women" in the name or focus yields a list of 20 committees. Of that number, 12 have focused on women's education and employment in the science and engineering workforce (the remaining committees looked at health-related issues of women and/or children, violence against women, and women's issues in transportation) (Table 9.1).

Some of these committees were specifically organized by the NAE (indicated in the table), while others (such as the still-active Committee on Women in Science, Engineering and Medicine (National Academies of Sciences, Engineering, and Medicine, n.d.)), were included the NAE from the beginning but were run by the National Research Council (NRC). Members of these committees included many women who were employed in academia at the time, as well as some women who are/were NAE, National Academy of Sciences (NAS), and/or National Academy of Medicine (NAM) members at the time of their membership. The number of academic women on each committee varies by the focus of the committee, so the total

J. P. Carpenter (✉)
Campbell University, Buies Creek, NC, USA

J. P. Carpenter (ed.), *Women in Academia*, Women in Engineering and Science,
https://doi.org/10.1007/978-3-031-97938-5_9

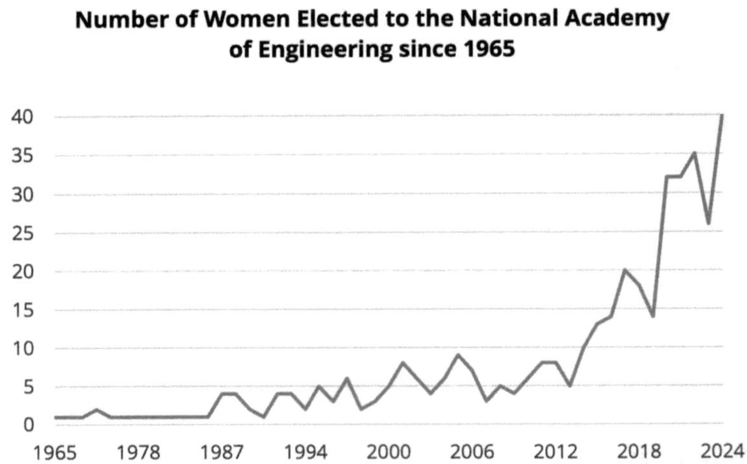

Fig. 9.1 Number of women elected to the National Academy of Engineering since 1965 (Source: Society of Women Engineers (2024). *National Academy of Engineering*. Retrieved February 19, 2025, from https://swe.org/research/2024/national-academy-of-engineering/)

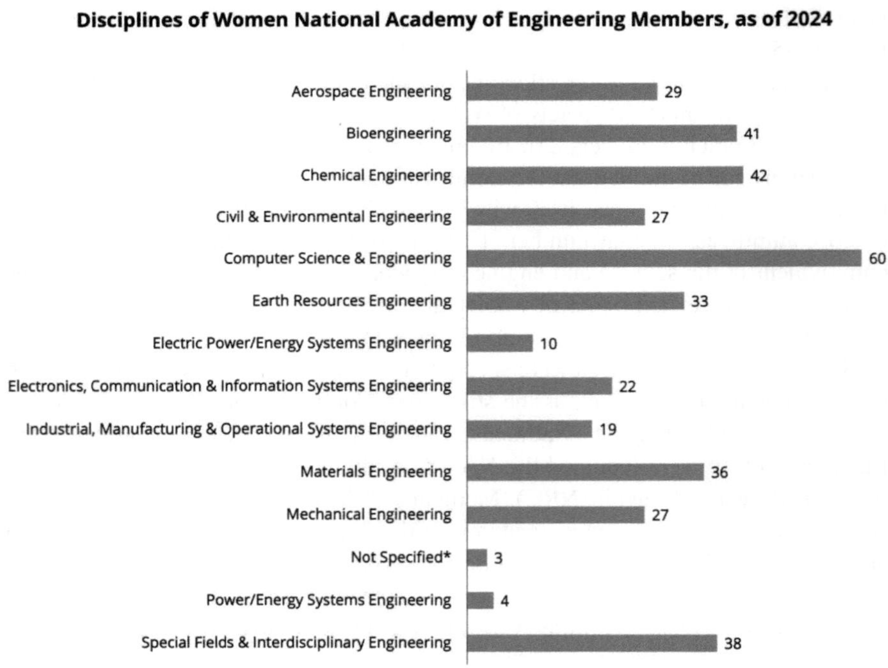

Fig. 9.2 Disciplines of women members of National Academy of Engineering, as of 2024 (Source: Society of Women Engineers (2024). *National Academy of Engineering*. Retrieved February 19, 2025, from https://swe.org/research/2024/national-academy-of-engineering/)

Table 9.1 Committees of the National Academy of Engineering and the National Research Council focused on issues pertaining to women in engineering academia and the engineering workforce

Committee name	Year organized	Year completed	Number of academic women members[a]	Number of women members[a]
Committee on Education and Employment of Women in Science and Engineering	(No date listed)	1983	6 (out of 18)	16 (out of 18)
Committee on Women's Employment and Related Social Issues	1981	1991	N/A	N/A
Panel on Technology and Women's Employment	1984	1986	N/A	N/A
Task Force on Women's Involvement in TRB and Transportation	1986	1990	N/A	N/A
Planning Group to Assess Possible OSEP Initiatives to Increase the Participation of Women in Science and Engineering	1988	1988	N/A	N/A
Committee on Women in Science, Engineering and Medicine	1990	(Active)	49 (out of 111)	82 (out of 111)
Panel on Barriers Faced by Developing Country Women Entering Professions in Science and Technology	1994	1995	N/A	N/A
Celebration of Women in Engineering Organizing Committee (NAE)	1997	1999	N/A	N/A
Forum on Diversity in the Engineering Workforce (NAE)	1999	2004	7 (out of 62)	28 (out of 62)
Committee on the Diversity of the Engineering Workforce (NAE)	1999	2005	6 (out of 25)	14 (out of 25)
Committee on Women in Academic Science and Engineering: A Guide to Maximizing Their Potential	2005	2006	14 (out of 19)	18 (out of 19)
Convocation on Capitalizing on the Diversity of the Science and Engineering Workforce in Industry	2008	(No end date is listed but all member terms ended in 2010)	0 (out of 12)	5 (out of 12)

[a] Membership lists shown for those committees to which we had such information. Some committee members served multiple terms, in which case they were counted each time they served. Committee member affiliation (academic, nonacademic) is based on their affiliation at the time of their service on the committee

number of women on the committee is also listed (for those committees for which membership lists were available) for comparison.

9.2 Turn of the Twenty-First Century

Initiatives focused on women began to gain more traction at the NAE in the late 1990s and early 2000s under then NAE President Bill Wulf. In 1998, he initiated the Celebration of Women in Engineering Steering Committee. They tackled two goals—one, create a website with information on pursuing an engineering career, and two, convene a summit to create a national action plan that engaged leaders from every sector impacting engineering and engineering education. Later that year, the December 1998 issue of *The Bridge* contained Wulf's now well-known article on *Diversity in Engineering*, where he noted: "Every time an engineering problem is approached with a pale, male design team, it may be difficult to find the best solution, understand the design options, or know how to evaluate the constraints" (Wulf, 1998a).

"The Celebration of Women in Engineering Steering Committee spent more than a year planning the Summit on Women in Engineering, which was held on May 17–18, 1999. The Summit brought together 175 of the nation's leaders from industry, academia, federal and elected government, professional engineering societies, media, foundations, and education and outreach organizations" (National Academy of Engineering, 2002). We were not able to locate a complete list of participants or formal proceedings for this Summit, but 15 presentations were compiled into a consensus study entitled "Diversity in Engineering: Managing the Workforce of the Future" (National Academy of Engineering, 2022). We also located copies of speeches from the event by Neal Lane, Assistant to the President for Science and Technology (in the Clinton Administration) (Lane, 1999), and an edited version of the speech given by Nance Ramsey, Former Legislative Director to Senator John Kerry (Ramsey, 1999). In his speech, Lane noted, among other things, that underrepresentation of women in science and engineering was a problem because STEM fields offer rewarding career opportunities that should be open to all; diversity of scientists and engineers assures better scientific and technical results; and we as a nation need people with the best minds and skills, no matter their race, gender, or ethnicity. He went on to state that:

1. The nation needed more support for undergraduate and graduate education including engagement from industry.
2. We needed to do more research to understand the challenge and barriers to participation of women in the science, engineering, and technology workforce.
3. "I particularly want to encourage agencies to think about diversity as a part of everything they do, not just special programs" (Ramsey, 1999).

Ramsey, in the article based on her speech, called out women pioneers in computing; male advocates and allies who have encouraged and welcomed women; STEM students and their achievements; and lastly, institutions and organizations that support women's success. She also noted demographic issues that we see clearly today—the fact that women outnumber men in college, reliance on "foreign nationals" to fill engineering workforce needs, and workforce shortages due to decreasing birthrates and retirement of baby boomers. She closed by noting while diversifying engineering has proved to be as challenging a problem as some of the greatest engineering achievements in civilization, if engineers can go to the moon, we should be able to solve this problem, too: "If we can put women into business, law, and medicine at levels nearly equal to men, why can't we do it in engineering? If engineers can put people on the moon and robots on Mars, surely they can figure out how to put women in engineering" (Ramsey, 1999).

After the Summit in May 1999, the subsequent June 1999 edition of *The Bridge* focused on Women in Engineering (including an edited copy of Ramsey's speech at the Summit), with an editorial by Wulf and NAE member E. Gail de Planque, where they again note the aging engineering workforce, declining numbers of engineering majors, and lack of diversity in engineering: "For the United States to remain competitive in a global, technological society, the country as a whole must take serious steps to create a diverse, well-trained, and multicultural workforce" (Wulf & De Planque, 1999).

Just one month later, in July 1999, Wulf testified at the Commission on the Advancement of Women and Minorities in Science, Engineering, and Technology Development, focusing on the status of women in engineering, the success (and lack thereof) to date in attracting women to the field, and the issues behind our collective progress which, despite the 25-year gap, look similar in many ways to the status and issues today. The work of the Celebration of Women in Engineering Steering Committee is highlighted in his report on the NAE's work, including creation of a now-defunct website (National Academy of Engineering, n.d.-b).

Other similarly focused committees (which do not specifically include the word women in their title) have existed at the NAE, such as the Committee to Advance Participation in Engineering or CAPE (founded in 2020 as the Committee on Racial Justice and Equity by NAE President John L. Anderson and renamed in 2025 "to better reflect its work") (National Academy of Engineering, n.d.-b). CAPE focuses on supporting the NAE mission "to advance the welfare and prosperity of the nation" by addressing participation in the NAE, the engineering profession, and the nation (National Academy of Engineering, n.d.-c). Reports exist from many of these committees. The National Academies Press has a searchable database of these and other similar publications (National Academies Press, 2025).

A few years after the flurry of activity in 1998–1999, the NAE launched EngineerGirl (in 2002) as part of a multipronged effort to move the needle on the engineering diversity gender gap (EngineerGirl, 2025). In particular, as its name suggests, EngineerGirl strives to reach young women in K-12, spark their interest in engineering, empower their resolve, and help them connect with the larger community of women in engineering. Their website contains many resources for encouraging girls and hands-on project ideas for students and teachers, along with an "Ask an Engineer" forum (EngineerGirl, n.d.-a) where students can submit questions for an engineer to answer. EngineerGirl sponsors an annual writing contest about the impact of engineering on our world, as well as an ambassador program that trains and funds young women to lead engineering projects in their own communities. As for impact, the writing contest alone has engaged over 1000 students (National Academies of Sciences, Engineering, and Medicine, 2022). EngineerGirl Ambassadors have conducted interviews with women engineers and NAE members, such as Maria Klawe, former President of Harvey Mudd College; Christina Zavaleta, Assistant Professor of Biomedical Engineering at the University of Southern California; Fiona Doyle, Professor of Materials Science and Engineering at the University of California, Berkeley; incoming NAE President Tsu-Jae King Liu, then Dean of Engineering at the University of California, Berkeley; and Lili Deligianni, Founder of sense4me Inc. and Adjunct Professor at the University of Illinois at Urbana-Champaign (EngineerGirl, n.d.-b).

In a 2022 NAE article on the 20th anniversary of EngineerGirl, it was noted, "A recent National Academies report finds that women who pursue engineering and other STEM careers face a range of barriers, including pay inequity, higher teaching loads, and sexual harassment. However, the roadblocks to pursuing a career in engineering commonly appear much earlier in the STEM pathway, said Simil Raghavan, a program officer at the National Academy of Engineering (NAE). 'All too often, girls are discouraged from pursuing their interest in the sciences, and start dropping away from STEM paths around middle school.'" (National Academies of Sciences, Engineering, and Medicine, 2022).

9.3 Recent History

Starting in 2019, the NAE Annual Meeting began hosting a Women of the NAE member-led event and lunch. Similar in focus to that of CAPE, the NAE website describes the purpose of the event as building "on the recommendations that came out of the June meeting, and to identify concrete goals, objectives, timelines, and volunteers to move from conversation to action, in support of the NAE objectives to increase diversity of nominations, including the diversity of successful nominations" (National Academy of Engineering, n.d.-d).

Most recently, the Fall 2024 issue of *The Bridge* focused on engineering a diverse future, with an article by the current NAE President John Anderson, "The Power of Diversity, Equity, and Inclusion in Engineering," where he notes, among other things, that the issue "addresses DEI in the engineering profession and articulates why progress in all three areas is critical for the advancement of engineering and technology in the United States." In his article entitled "The Uncapped Potential: Engineering an Opportunity of a Lifetime," Norm Augustine notes that "there are self-imposed problems limiting engineering in America, one of which is the constraint the nation, usually short of engineers, places on itself by failing to attract to the field more talent from the rich variety of backgrounds in its populace." In their article "The NASEM Diversity Science Report: Going Beyond Mere Participation

Numbers," Gilda A. Barabino and Susan T. Fiske, co-chairs of the consensus study that produced this report, summarize their evidence-based recommendations, such as looking at new education pathways for individuals in STEM and evaluating culture and anti-racism overall, with an eye toward improving the sense of belonging, interaction with in-group peers, mentoring opportunities, and examining norms of inclusion. Megan Smith, Wanda Sigur, and Puneet Ahira in their article "Endless Talent is the American Dream: A Draft Blueprint for Realizing the Full National Potential" lay out detailed steps and methods which interested companies and groups can use to realize diversity and inclusion (National Academy of Engineering, 2024).

9.4 Conclusion

The NAE has admitted women members since almost its very beginning. Early numbers were small and progress slow, reflecting in part the roles women were allowed to hold within engineering in society at the time and public interest in and support for women in engineering roles. Spurred by the efforts launched under then NAE president William A. Wulf, the last 25 years have seen significant progress in diversifying NAE membership and intentional focus devoted to trying to solve the engineering education and workforce diversity problem. At this writing, we stand on the cusp of the predicted demographic decline in the USA, along with sweeping efforts at both the state and federal levels to reverse initiatives and legislation which have supported efforts to diversify engineering. Only time will reveal the impact of this wave.

9.5 Interviews with Academic Women NAE Members

We interviewed five academic women members of the NAE whose election has spanned roughly the last 25 years and asked them several questions to gain insight into the changes they have seen regarding career opportunities and climate for women engineering faculty, how people reacted to their being elected to the NAE, and how it has impacted their career, as well as what advice they have for women and girls today who want to pursue an academic career in engineering and computing.

9.5.1 Dr. Linda M. Abriola

Joan Wernig and E. Paul Sorensen Professor of Engineering

Brown University

Elected 2003, Civil and Environmental Engineering, Special Fields and Interdisciplinary Engineering

9.5.1.1 Thinking Back Over Your Career, How Have Things Changed for Academic Women in Engineering and Science?

There have been a lot of changes. When I started my career, there were very few women in engineering and science, maybe more in the life sciences, but very few in engineering and physical sciences. I was the first woman in my department when I was hired at Michigan. The university administration made a concerted effort at the time to bring in more women faculty across the engineering college, and the administration was generally supportive of these women. Unfortunately, many women ended up leaving the university (and even academia) within a few years, in part because there were huge service demands and no family-friendly policies. You had to be much better than the average male faculty member to succeed. Today, we see a lot of young women entering the profession. It is still difficult to obtain tenure and survive, but there are a lot more policies on childcare and there is more shared responsibility for family life. All of this has improved the situation for women in the academy. A lot more women have been elected to the NAE and NAS. People are more cognizant now of the service burden we assigned to women in the past, and they have worked to address that, so there are more women in senior faculty positions and leadership roles now. Early in my career, I had a very large service role but today, with more women in the academy and the realization that you do not have to have a "token woman" on each committee, that has improved. The other important point is that I had no role models when I began my career. Today, we can see women winning the Nobel Prize and being successful in academia, so it is much easier to visualize ourselves being successful in the role.

9.5.1.2 What Issues Still Exist that We Need to Address?

I still think women who succeed have to be a little bit superhuman. I think young women getting PhDs are seeing that, and they are choosing not to enter academia. This perception of the challenges of an academic career is still a factor in the leaky pipeline. How we address this is a more complicated question. The tenure timeline coinciding with childbearing years is still an issue, as is the "one-size-fits-all mentality" that is pervasive in academia. Individuals contribute in different ways to their institutions and the profession throughout their careers, and I believe that there needs to be more consideration of this in tenure and promotion decisions. It is hard, however, to generalize. At specific institutions, it is easier to identify local issues that we need to address. For example, I believe that if we can more greatly value interdisciplinarity and cross-disciplinary scholarship, this will help women succeed. It may be that at many institutions, it is already greatly valued, but at many elite institutions, this is still not the case. In general, it is much more difficult to be successful if you are working across disciplines or with collaborators across disciplines. Many women I have known prefer that model of scholarship. The tight timeline for tenure, the need for aggressive self-promotion, and the expectation of disciplinary scholarship as first author all make it more difficult for women to succeed.

9.5.1.3 How Did People Respond to Your Election to the National Academy?

My background is such that I was the first woman in the family to go to college. My parents, my aunt and uncle, and my brother and his family showed their support and came to the induction ceremony, but they did not really understand the honor that I was receiving. They were impressed at the ceremony when I was standing on the stage with all of those "important people" and were happy for me. With respect to my academic colleagues, there was a little whiff, not a lot, of "you are a woman, and they are trying to get more women in the academy." I was also younger than the average male member when I was inducted, and I had a number of colleagues comment that I was "so young," as if that also indicated a preference related to my gender. So, I was left with a vague unease and that feeling of unearned success. Later, I saw the election numbers and learned that the National Academy was indeed trying to increase nominations of women to the academy. However, although there were more women nominated than in prior years, very few were actually elected. Then when I walked into the civil engineering section meeting after my induction, I found that I was the only woman in the room. I asked if there were any other women in the civil engineering section of the academy, and no one seemed to know. It turns out that there was one other elected woman in the section who was not in academia, so I guess I was the first elected academic woman in my section. I had this feeling at the time that I was back where I started, in terms of being the only woman again, and right away, I was asked to do a lot of service on Academy committees. However, as with academia, much has changed in the National Academy in the intervening years. I love the Academy, and I believe that it is doing a lot of great work.

9.5.1.4 In What Ways Has It Impacted Your Career?

It made a big difference on my marketability as a faculty member. I was hired as dean of engineering at Tufts around the same time as my induction. I cannot really say if it played a role. People at Michigan knew that I was under consideration, but I did not know it. Even before I was inducted, I served on NRC study committees, which was really valuable because it enlarged my vision of the field. I do not know if my election directly impacted my research, but it enhanced my worldview and made me more appreciative of the scope and impact of engineering. I have met so many extraordinary people in the Academy, people of such achievement. It has been for the most part a really welcoming community. It gave me a better perspective on academia and business and how they can work together to solve problems and how engineering and science can impact public policy. I did not have as much of an appreciation of that before. The women in the Academy now meet as a group at every annual meeting and talk about what we can do to enhance the election and engagement of those underrepresented in the academy.

9.5.1.5 What Advice Do You Have for Women and Girls Today Who Are or Want to Pursue an Academic Career in Engineering or Computer Science?

Despite the challenges, I still believe that academia is an amazingly rewarding career. It allows you to shape your life the way you want it. You can have connections with people, the opportunity to educate the next generation, and the opportunity to choose what you want to work on. I think of academia as very entrepreneurial. My dad started and owned a construction firm, and I think it is very much like that. To fashion your own path and work on your own projects and work closely mentoring people, to be able to do many things and make it all part of your career, it is very rewarding. Having the flexibility to make your own path is wonderful. You also have a lot of flexibility in shaping your career as you get older. This is what I also tell my students. I tell them not to make the mistakes I made—to protect themselves from too much service and to speak up when they think they are not being treated fairly. I also have tried to serve as a mentor and be protective and supportive of those women faculty coming up "in the ranks" behind me. I review their CVs, make suggestions, etc., to help them succeed. I was quite isolated starting out in my career and want to help alleviate this isolation for others. It is wonderful to see the successes of the next generation. In this profession, you touch a lot of people, and then they go out and do much better things than you could ever do. It is a real joy.

9.5.2 Dr. Leah Jamieson

Ransburg Distinguished Professor of Electrical and Computer Engineering, Professor of Engineering Education (by courtesy), and John A. Edwardson Dean Emerita of Engineering

Purdue University

Elected 2005, Special Fields and Interdisciplinary Engineering, Computer Science and Engineering

9.5.2.1 Thinking Back Over Your Career, How Have Things Changed for Academic Women in Engineering and Science?

The most obvious change is in the numbers: they have changed dramatically for academic women. When I started as a professor at Purdue, I was the second woman faculty member in electrical engineering, out of a total of about 70 faculty. More broadly, there were only three women faculty across all of engineering at Purdue at that time. So, as a woman, you were automatically an outlier and, in some ways, a new experience for your faculty colleagues. The numbers were small enough that with surveys and data collection, women's responses fell in the "too small to report category," to protect confidential information. The number of women students has obviously grown as well, so you see more women in your classes—but not so many that you can stop thinking about team dynamics when someone is the only "other" on a team. Purdue had the first Women in Engineering program in the country, so for over 50 years, there has been a support network for women students and, for those who wanted it, a community of women. For many years, the women faculty mostly formed our own support network and community, but we also had support from some inspired (and inspiring) heads and deans.

A phenomenon that has not gone away is that the service loads for women (and faculty from other minoritized racial and ethnic groups) were (and remain) high. People wanted us on every committee, to add the "women's point of view," and there were a lot more committees than women faculty. This meant that the ways women faculty were being asked to spend their time were different. There is no question that the committee load for women faculty was higher than that of, say, white male faculty, not to mention the burden of representing "all" women. From a career standpoint, this disproportionate service load needed to be recognized. Another area where thinking has changed (and is still changing) is childcare. Early on in my career, it was definitely a women's issue. I would not say that thinking is behind us, but it is changing over time.

I cannot help but remember markers spotlighting why it was (is?) imperative that the thinking about women evolve over time. I once presented a promotion and tenure case for a woman faculty member in electrical engineering whose research was in an area in which many of the outstanding leaders were women. As a result, all of her letters, which were wonderful, were from women. Someone on the committee commented that this was odd, in a manner that suggested that it weakened the case. I wish I had had the presence of mind to point out that no one had commented on the cases where all of the letter writers for male candidates had been men.

I find that, even with greater numbers of women faculty, community (and community building) is still important, although it sometimes takes different forms and is probably more nuanced. I was elected to the National Academy of Engineering in 2005, the same year two of my colleagues and I were awarded the NAE's Gordon Prize for Innovation in Engineering and Technology Education. Purdue's practice has been that if you are elected to the NAE and are not yet a distinguished professor, you are typically put on an expedited path to appointment as a distinguished professor. I became a distinguished professor in 2006, which was also the year I became dean, and the year I was President-Elect of IEEE, so my life was on the verge of changing with or without NAE membership. For eight years, I was the first—and only—woman to hold a distinguished professorship in all of engineering. Finally, in 2014, there were two of us. Ten years after that, in 2023, three women faculty from across engineering were named distinguished professor on the same day. We went from two to five overnight. We celebrated!

So, although things have changed, there are still some vestiges of situations where women are rare. It seems to track seniority and age. I have seen that used as an excuse for why there are fewer women in some settings. But women have been in engineering and computer science in sufficient numbers for long enough that people should not be making excuses that there are no enough women ready for recognition at the highest levels. For example, starting in 2015, the NAE has seen a dramatic increase in the number of women elected to membership. We celebrate that! But I still hear some of the stock phrases that were common in past decades. A little over a year ago, in response to a suggestion about making sure that women were being considered for a high-level position in an organization, one of the men in the meeting immediately reacted: "We don't want to lower our standards." As if that were the first thing that comes to mind when thinking about having a woman leader. I do not hear it as often as I used to, but it is still out there.

9.5.2.2 What Issues Still Exist that We Need to Address?

I would like to see us collectively recognize that careers and leadership are not gendered and that issues such as work-life balance are not women's issues. It is getting better, but we are not there yet. There is still evidence of stereotypes about men's jobs and women's jobs. And the challenges faced by arguably the senior-most women in academia are sometimes formidable. In the past year, we have watched women university presidents resign under fierce questioning about campus protests over the Israel-Hamas war, giving every appearance of a campaign to drive out these leaders. This is bad for everyone because eventually people will say, "I do not want that job." That will be a loss for all of us.

9.5.2.3 How Did People Respond to Your Election to the National Academy?

Purdue was thrilled. My family was thrilled. My community of academic friends in technical fields across the country was also thrilled, but there was some puzzlement about my election. My NAE citation talks about innovations in integrating engineering education and community service—not your typical NAE citation. We started the Engineering Projects in Community Service (EPICS) program at Purdue in 1995. I was elected to the NAE 10 years later. Colleagues at the time asked me if EPICS was taking time away from my research. When I said yes, the frequent response was "that's too bad." But it did not feel too bad to me. I was one of the early members elected to the Academy on the basis of contributions to engineering education. We would see citations that noted both the member's technical contributions and their contributions to education in their field, but this idea that someone's election could be based primarily on engineering education was different. Colleagues did not know what to make of it, because it was not about my research. I think that the NAE has done a better job of figuring it out. The committees to which I have been appointed and on which I have been invited or have chosen to participate have included NAE committees, but have also spanned the National Academies of

Sciences, Engineering, and Medicine (NASEM). For example, I served on the 2008 NAE Changing the Conversation steering committee, which was squarely centered on engineering and the public understanding of engineering. I am currently one of the three co-chairs of the member-led Women of the NAE Committee, focused on community building and support. But I have also served on the steering committee that took an early look at the impact of COVID-19 on women in academia across the disciplines represented by all three Academies and on a steering committee focused on barriers and opportunities to completing 2- and 4-year degrees across all of STEM. I have chaired the Academies' committee that selects recipients of the Schmidt Awards for Excellence in Science (and Medicine and Engineering) Communication, and I serve on the advisory committee for the National Academies Division on Behavioral and Social Sciences and Education. I am an NAE member with a broad portfolio, which is perhaps a good description of me more generally.

9.5.2.4 In What Ways Has It Impacted Your Career?

Election to the National Academy is such a definitive statement of your success in academic circles. It is a respected credential, a stamp of approval. It has opened doors and led to speaking invitations. It has also given me the opportunity to be a part of NAE and NASEM studies, to add my voice on critical issues and expand my perspectives in ways that affect my thinking every day. And interestingly, since I stepped down as the dean of engineering at Purdue in 2017, my membership in the NAE has also, to a large extent, shaped my service activities at Purdue. I have served on and chaired awards and recognition committees at the school, college, and university levels. It has also made mentoring and nominating others for recognition a substantial part of my academic life.

9.5.2.5 What Advice Do You Have for Women and Girls Today Who Are or Want to Pursue an Academic Career in Engineering or Computer Science?

For girls, look for opportunities to start doing, learning, and finding out what the various fields in engineering and computer science involve. Go to summer camps, weekend programs, local museums, and national science centers. There is a lot to love about engineering and STEM in general, but it is collectively a huge field, and it works best when you find your passion. I advise and coach design teams of students from engineering, as well as from fields all across campus, on designing and building engineering- and technology-based solutions in partnership with nonprofit community agencies. It gives me both energy and optimism. Look for ways to explore not only the technical side of engineering, but also the ways in which engineers work—e.g., that it is a team sport—and all the ways that engineers make a difference in the world.

The same principle applies to women looking to pursue an academic career. Look for or create your own opportunities to gain experience in the range of roles that are a part of being an engineer in academia. Find ways to understand what you are excited about and know why you are excited about it. It gives you the determination that any challenging—and rewarding—career will require.

9.5.2.6 Any Other Thoughts You Would Like to Share?

I continue to love seeing the successes of women in engineering and computer science. When I first started at Purdue as the second woman in the field, senior faculty would come by my office and say that the women in their classes were the best students in the class. My response was "that's too bad. They shouldn't have to be." We were fine with having average men, but if a woman was going to do engineering, then they had to be the best. I think I still carry this with me. Finding ways to make any of the professions related to engineering open, so there are no barriers from family, classmates, teachers, employers, and others, continues to be important. If you are excited about engineering or computing, there should be a path for you to do it. And we need you. As a society and as problem-identifiers and problem-solvers, the quality of our solutions will depend on the diversity of both life experiences and thinking of the people around our table. I still have a magnet created at the time of the release of the NAE's Changing the Conversation report. It says "Engineering Challenges: Because Dreams Need Doing." It inspires me every day. Find your inspiration.

9.5.3 Dr. Frances S. Ligler

Eppright Chair and Professor of Biomedical Engineering
Texas A&M University Distinguished Professor
US Navy Senior Scientist for Biosensors and Biomaterials (ret.)
Elected 2005, Bioengineering, Special Fields and Interdisciplinary Engineering

9.5.3.1 Thinking Back Over Your Career, How Have Things Changed for Academic Women in Engineering and Science?

When I first came through, there was a tremendous amount of discrimination. It was assumed that women were responsible for taking care of the family so they would not have time to do anything special. I had three job offers vetoed by a major medical school because I was a female PhD. I also had bad experiences being interviewed for academic jobs after being a postdoc, so I went to industry. I spent 5 years in industry and 28 years in government before coming back to academia about 12 years ago. The big difference back then was that in industry, nobody cared what

I looked like; they only cared about what I could do. It was refreshing. It is not so bad in academia now. The fact that there are more women around has made a huge difference. My daughter had some issues going into the biotech patent business, but everything is more subtle now. My boss at NRL once said that he would have hired a qualified woman if he could find her, but that is no longer true. You do not have to be twice as good any more to get a job. You can spend more time on the work and less time trying to prove yourself.

9.5.3.2 What Issues Still Exist that We Need to Address?

I am not so worried about male versus female in engineering as in the past. We need to work on the identity of an engineer for women to be able to see themselves in that role. Some of that comes from seeing other senior engineers being women so that they can identify with them. Understanding "what an engineer is" is a moving target and that is good. The idea of people doing dense problems by themselves is gone. When people see engineering as a team sport and working on relevant topics, it will be much easier for women to see themselves as engineers. Anita Jones once said that engineering is a hands-on sport.

9.5.3.3 How Did People Respond to Your Election to the National Academy?

I think everyone was excited for me. It was unexpected. I was only the third woman in Section Two of the Academy, so it was kind of a new thing. Everyone I work with was very excited and pleased because it was kudos for what we all had done, not just me.

9.5.3.4 In What Ways Has It Impacted Your Career?

For me personally, it opened a lot of doors and not just in my field. I have gotten to meet such terrific people. It has given me an opportunity to do work and contribute to reaching goals well beyond my own expertise and the projects I work on, much more on a national and international level than I would have had the chance to do otherwise. I have learned an awful lot.

One more thing that has come about because of who I have gotten to know through the Academy is my involvement with the National Inventors Hall of Fame. The best part of that is going every summer to Camp Invention for elementary kids to show them that they are inventors. That has been an amazing part of my life for the last few years. The kids are fabulous. They are not inhibited by what they do not know. It is a great lesson for all of us.

9.5.3.5 What Advice Do You Have for Women and Girls Today Who Are or Want to Pursue an Academic Career in Engineering or Computer Science?

For women in junior high and high school, I would tell them to get involved in a team sport or some equivalent team activity because learning to work with other people as part of a team is so critical to succeeding in science and engineering. There are two parts of this that are really important. It is not just trying to articulate what you want to say in a way that other people will understand, but it is also developing enough empathy to understand how others will perceive your words. This can be tricky, but if you want your team to succeed, you have to understand two-way communication.

Be very aware of what problems you care about, what you want to work on, and what problems you are passionate about solving, because you are going to need that kind of motivation to teach and establish a research program. Collaborate as much and as widely as you can, not just within your institution but outside as well. There is no way you can solve a major problem with what is just in your own head. You will need collaborators and criticism and competitors. You have to do the work in concert with other people if you want to succeed.

Never be afraid to listen to others—as many sources as you can, up and down—you do not have to follow it if it is not good advice. Academia is an exciting and a wonderful life. You will be a citizen of the world in no time. I have lived as much in one lifetime as most people do in three because of all the doors that have opened for me.

9.5.4 Professor Tsu-Jae King Liu

Dean and Roy W. Carlson Professor of Engineering
University of California, Berkeley
Elected 2017, Electronics, Communication and Information Systems

9.5.4.1 Thinking Back Over Your Career, How Have Things Changed for Academic Women in Engineering?

First of all, engineering is a broad field, and there are differences among the various subfields, but overall, there are more women in engineering now than when I was a student. In all my years as a student, I only took one class from a female engineering professor. Today, when I meet with deans, we observe that we have all seen growth in the number of women in the professoriate. We pursued engineering because we wanted to make a difference in people's lives, to benefit people and society. But if engineers are going to do so, they need to understand the challenges of all people in society. One way to ensure that engineering teams have that broad understanding is to have the wide representation of society reflected on those teams. It is also important that the diversity of people in engineering grows as people in society become more and more dependent on technology and other tools that engineers have innovated and designed. So, while we have made a good first step, we need to continue to see more women pursuing engineering.

There was an NAE report written 20 years ago, in 2005, entitled "Educating the Engineer of 2020." This report noted that any country weak in technological literacy will have an increasingly difficult time in competing in the economy of the twenty-first century. So, we need to make sure that people in the USA have some understanding of technology and some degree of technological literacy, or we will not thrive as individuals or as a country. This report also noted that engineering educators should participate in a coordinated effort to promote this public understanding of both engineering and technology literacy. If we in academia want students to learn and be inspired, then those students need mentors and teachers to whom they can relate. So, our faculty need to come from a wider range of backgrounds and perspectives. Collaboration is another important characteristic of engineering teams. People must be able to relate well to one another in order to collaborate effectively. In general, women are perhaps culturally socialized more toward collaboration (than competition). In higher education, we should be fostering more collaboration among educators themselves if we want to teach out students to do so.

9.5.4.2 What Issues Still Exist that We Need to Address?

The average math achievement score has been steadily declining for all demographic groups in the USA in recent years, which is a worrisome trend since a solid foundation in math and science is needed to succeed in engineering. Clearly, this is an issue we need to address. Helping students understand how and why math is useful in engineering and other fields may create more interest in learning math.

9.5.4.3 How Did People Respond to Your Election to the National Academy?

In my own community, people were happy and congratulatory. However, in general in the USA, engineering is not well understood and therefore not as appreciated as it should be. Therefore, many people do not know about the National Academy of Engineering. Election to the NAE is one way to raise its visibility and recognize people who have made important contribution to the field of engineering. It is good for the NAE when a broader swath of the campus understands that the NAE works to improve engineering in this country. We have a National Science Foundation and a National Science Board (5 of its 25 members are engineers), but no National Engineering Foundation or National Engineering Board. The PCAST Committee does not include engineering. So fundamentally in the USA, engineering is sometimes invisible to people. They know perhaps that they depend on what engineers build but they do not really know what engineers do. They have some idea what science is but do not know enough to attribute discoveries to engineering (versus science). So, this lack of visibility for engineering is to our detriment. We do not have enough talent going into engineering in the USA which makes it difficult for us as a country to stay competitive in a technologically driven and dependent world.

9.5.4.4 In What Ways Has It Impacted Your Career?

It has certainly helped it! Election to the NAE is the highest honor in this country that can be afforded to an engineering professional. And it is always good to have some affirmation that your work has had impact in the field. As a woman, there is a natural tendency toward implicit bias (now we see it explicitly since people are more bold in speaking out against DEI initiatives). When a woman is appointed to a position, there is the tendency to ask if she is a diversity hire. Being a member of the NAE has helped to counteract that bias and reinforce that I was selected because I am qualified. It weakens your position as a leader if people do not acknowledge that you are qualified. I remember that whenever female students won a scholarship or fellowship male students would say that it was easier for them to win because they were a woman. You hear it is less so today than when I started as a professor, but that narrative is still there.

9.5.4.5 What Advice Do You Have for Women and Girls Today Who Are Pursuing or Want to Pursue an Academic Career in Engineering or Computer Science?

We need more women in engineering. Engineering is about developing solutions that benefit all people in society, so ideally 50% of our educators and researchers in academia need to be women. A career in engineering is also a wonderful way to have an impact. The measure of success as an engineer is really the measure of your impact. Academia is a good career path because you get to teach thousands of students over your career and your impact is magnified through them. In addition, the research you do advances knowledge and your field. It gets presented publicly so that it can benefit other academic researchers and industries. All of these opportunities for impact make it a very rewarding profession. Society is becoming more and more dependent on technology, so to be sure it does not have negative consequences we need both women and men to design these innovations. We need to educate women and inspire them. Female students need role models to persist in engineering. The graduation rate overall in engineering is less than 40%, so how do we reduce this leakage in this pipeline? We need to change the culture and transform it to be welcoming, inclusive, collaborative, and supportive so that everyone can succeed.

9.5.4.6 Any Other Thoughts You Would Like to Share?

Women in academia influence students, and it is just as important for women in academia to inspire men as well as women. Men, if they have positive experiences with a woman mentor, can become allies and advocates for women. I had a former student tell me that he does actually speak up for and support women now. His attitude has changed since he was a student. We should keep this in mind.

9.5.5 *Dr. Gilda A. Barabino*

President
Olin College of Engineering
Elected 2019, Bioengineering, Chemical

9.5.5.1 Thinking Back Over Your Career, How Have Things Changed for Academic Women in Engineering and Science?

Looking back, you cannot help but look at the representation of women in these fields and how it has changed over time, but also the changes in the culture, the quality of interactions, and what the environment is like for women. You can see over time that the situation where women who were "solo" or the only one in their unit is starting to change a bit. You can also see it at various levels. At first, there were more women entering fields at the student level, then some parity at the assistant professor level, etc. But the parity decreases as you move up and as you look at leadership roles.

But it is not enough to look at numbers. You also have to look at changes in the environment that support women's success. I have been studying women faculty for some time. Early on, we were focused on "fix the women." But systems, cultures, and environments have to change if we want to see a difference. And some of the cultural shifts have been because of the presence of more women and the perspectives women bring to the table, particularly when they have the opportunity to lead.

9.5.5.2 What Issues Still Exist that We Need to Address?

The first issue that always comes to my mind is parity. Where are the equity gaps? I do not feel that women are afforded the same opportunities as others, and when they are, they are not given the same level of support. There is something to the concept of glass ceiling and glass cliff. We will entertain hiring a woman when everything is messed up, but when everything is going well, we tend to hire those who look like the people who have traditionally been in charge.

Despite the successes women have had so far, we are still dealing with societal perspectives that influence how women are viewed and treated. We have to position women for success. Women receive more scrutiny. They have more trouble corralling the support they need when things get tough. We find it easier to throw the women to the curb. Women also tend not to get the recognition they deserve. When you have an entity based on what you have achieved, something with an honorific component to it, you have an entity that is less likely to recognize and reward women. It is still harder for women to rise to the level of being nominated and selected. Women do the hard work and carry the weight on their shoulders, yet sometimes they are not the ones being applauded and getting the credit for their work. They do the work because they are trying to do what is right and because they are focused on getting that end result. Women are not good at tooting their own horn. It is hard for them to promote themselves. Women are told to be quiet and keep their head down.

The power dynamics have to shift, because who is in control matters, and they have not shifted enough to ensure the success of women and success overall. If the numbers of women are low and they are not part of the selection committee, it is harder to ensure that women are recognized. It matters!

9.5.5.3 How Did People Respond to Your Election to the National Academy?

There were those already in my corner who were jubilant. There were those who knew and respected me who thought that my recognition was overdue. There were also those who were jealous and envious, those who tried to make light of the accomplishment. I received more attention and letters of congratulation from outside my institution than I did inside.

When someone in the "in group" gets recognized, people will go out of their way to celebrate, lift them up, and point out their accomplishments. When we fail to do that for women, we are told that that is just the way it is, but it is not the case. It is just a way to keep us quiet. People do what they want for whom they want.

Research shows that recognition is more important than confidence or competence, especially when it comes from those considered of high stature. High status recognition has more to do with your career progression than your competence or your level of performance. There is so much in the social science literature that talks about what women experience and speaks to why they experience these things. Most institutions just do not have the appetite to do anything about it. If we want to make a difference, we women have to do it ourselves, build each other up, and notify each other of opportunities.

9.5.5.4 In What Ways Has It Impacted Your Career?

Being nominated and elected are impactful because of the recognition you receive. In your field, recognition particularly from your peers, those in your disciplinary area, makes a statement. Someone who may not know you or an outside organization or an entity that you are working with, they know what your recognition means, so it conveys to them expertise that can open doors.

One of the benefits has been getting to interact with others who have been elected to the Academy. When you are elected, you are entering another network you have not been in previously. What I like about the Academy is the expectation to serve once elected. The service expectation distinguishes it from other honorifics. If you take advantage of these opportunities to serve, it can advance your career, broaden your horizons, and increase your learning. It is a great vehicle and platform to be a part of, to learn and to serve. You can then apply what you learn and see the impact on your career overall.

Early on in my career, I felt like I needed to connect with others since I was the "only one," so I decided to serve. If you are volunteering to advance a project, initiative, or organization, it gives you a way to connect with others and get something done at the same time. I kept serving at higher levels. We learn more when we are out and about in other settings outside of our own institutions. It is when you broaden your world, your footprint, and not just in engineering. With AAAS, I am working with people across disciplines, institutions, geographies, goals, and objectives. Talk about enrichment! Bring this learning back to where you are. It distinguishes you from your peers because not everyone tries to lead or learn by serving. What you can gain for yourself and your institution is valuable. I do not see a substitute for serving.

9.5.5.5 What Advice Do You Have for Women and Girls Today Who Are Pursuing or Want to Pursue an Academic Career in Engineering or Computer Science?

I think my advice in general, as it applies all the way through your career, is to be open to new opportunities, new learning, and new ways of thinking. Be open and bold enough to try new things, even if it feels like a stretch. It takes a certain level of courage to be open and to act on wanting to learn. You have to be motivated to serve and want to help. But do not lose sight of your main goals; keep your eyes on the prize. To not lose sight, you have to draw on your courage and confidence, support yourself, and know who you are. Do not compromise. It is hard sometimes to fit in.

Another thing that has helped me through my career is to just be me. Do not try to be somebody else. If I have to deal with some consequences for doing so, at least I did not sell myself out. Young African American women will ask me how I am still standing. It helps to draw on support from others. You cannot do all the lifting of yourself by yourself. Draw on the support of family and peers who are trying to help each other. Sometimes you are only as good as the team you are working with. Sometimes we feel alone and isolated. Remember that we are not alone. We might have to do more to find those who are with us.

For African American women in particular who grow up in a culture that teaches them to be tough and take care of themselves and everyone else, it can be hard to ask for help. I think sometimes people buy into the stereotypes and sometimes they are not necessarily helpful. I speak out. Not everyone has the capacity to do so, but do what you can, even if it is a baby step. Maybe one day you point out an inequity in a conversation. Next time, you might take on something a little tougher. If we can all do what we have the capacity for, it will make a difference.

9.5.5.6 Any Other Thoughts You Would Like to Share?

John Slaughter says that we need opportunities to match talent. This is important because there are those who could be so much farther long if they had only had opportunities. Those of us how have had opportunities have an obligation to provide opportunities for others, especially when we know that the system is not set up to support people. We also need to broaden the criteria and definitions of success. Who gets to define those criteria? If you have different perspectives, the criteria are more likely to include those who have been left out. When you are in the system, then you have to take advantage of the opportunity to define the criteria and choose the next people to be included, so that we do not keep overlooking those who deserve to be recognized.

Cheryl Leggon, a social scientist at Georgia Tech, says, "who does science dictates who will do science." The same is true for awards. Who gets awards influences who will get awards. We need to remember this and be sure that we are being more fair and equitable in terms of who gets to participate in the first place since they will determine who gets to participate in the future.

Acknowledgements The views presented in this chapter are those of the author and not necessarily of the National Academy of Engineering or the National Academies of Sciences, Engineering, and Medicine. The author thanks Casey Gibson, Associate Program Officer at the National Academy of Engineering, for locating and suggesting resources and information pertaining to the history of women in the National Academy of Engineering.

References

EngineerGirl. (2025). Home. Retrieved from https://www.engineergirl.org. Accessed 22 Feb 2025

EngineerGirl. (n.d.-a). "Answers." Retrieved from https://www.engineergirl.org/366/Answers

EngineerGirl. (n.d.-b). EngineerGirl interviews. Retrieved from https://www.engineergirl.org/139476/EngineerGirl-Interviews

Lane, Neal. (May 18, 1999). *Remarks of the honorable neal lane, assistant to the president for science and technology, summit on women in engineering*. The White House. Retrieved from https://clintonwhitehouse3.archives.gov/WH/EOP/OSTP/html/998_27_2.html

National Academies of Sciences. (July, 2022). Engineering, and Medicine. Two Decades of EngineerGirl: Building a More Diverse Field. Retrieved from https://www.nationalacademies.org/news/2022/07/two-decades-of-engineergirl-building-a-more-diverse-field

National Academies of Sciences. (n.d.). Engineering, and Medicine. About CWSEM. Retrieved from https://www.nationalacademies.org/cwsem/about. Accessed 20 Feb 2025

National Academies Press. (2025). Browse by collection. Retrieved from https://nap.nationalacademies.org/collection/. Accessed 22 Feb 2025

National Academy of Engineering. (September 1, 2024). Fall bridge on engineering a diverse future. Retrieved from https://www.nae.edu/323119/Fall-Bridge-on-Engineering-a-Diverse-Future

National Academy of Engineering. (n.d.-a). Dr. Lillian Moller Gilbreth. Retrieved from https://www.nae.edu/30684/Dr-Lillian-Moller-Gilbreth. Accessed 20 Feb 2025

National Academy of Engineering. (n.d.-b). NAE President's Racial Justice and Equity Committee. Retrieved from https://www.nae.edu/238548/NAE-Presidents-Racial-Justice-and-Equity-Committee

National Academy of Engineering. (n.d.-c). Testimony to the Commission on the Advancement of Women and Minorities in Science, Engineering, and Technology Development. Retrieved from https://www.nae.edu/19579/31222/8436/8646/Testimonytothe Commissiononthe Advancement of Womenand MinoritiesinScienceEngineeringandTechnology Development

National Academy of Engineering. (n.d.-d). Women of the NAE Meeting: Lunch (Members Only). Retrieved from https://www.nae.edu/299331/Women-of-the-NAE-Meeting-Lunch-Members-only

National Institutes of Health (n.d.). The National Academy of Engineering. Retrieved from https://irp.nih.gov/about-us/honors/the-national-academy-of-engineering

Ramsey, N. (June 1, 1999). The future of women in engineering. In *The Bridge*. National Academy of Engineering. Retrieved from https://www.nae.edu/7495/TheFutureofWomeninEngineering

Society of Women Engineers. (2024a). National Academy of Engineering. Retrieved from https://swe.org/research/2024/national-academy-of-engineering/. Accessed 19 Feb 2025

Society of Women Engineers. (2024b). National Academy of Engineering List of Women Members. Retrieved from https://swe.org/wp-content/uploads/2016/08/National-Academy-of-Engineering-List-of-Women-Members.pdf

Wulf, W. A. (December 1, 1998a). Diversity in engineering. In *The Bridge*. National Academy of Engineering. Retrieved from https://www.nae.edu/7488/DiversityinEngineering

National Academy of Engineering. (2002). *Diversity in engineering: Managing the workforce of the future*. The National Academies Press. Retrieved from https://nap.nationalacademies.org/catalog/10377/diversity-in-engineering-managing-theworkforce-of-the-future

Wulf, W. A., & De Planque, E. G. (June 1, 1999). Women in engineering (editorial). In *The Bridge*. National Academy of Engineering. Retrieved from https://www.nae.edu/7457/WomeninEngineeringeditorial

Chapter 10
Her Deanship: Leadership in Academic Engineering as an African American Woman

Gilda A. Barabino

10.1 Introduction

> When I dare to be powerful—to use my strength in the service of my vision—then it becomes less and less important whether I am afraid.—Audre Lorde from *The Transformation of Silence into Language and Action.*

I recall the afternoon in the spring of 2013 while working in my faculty office; I received a call from a biomedical engineering colleague I had not yet met whose purpose in making the call was to convince me to enter the search for the dean of the Grove School of Engineering (GSOE) at the City College of New York (CCNY). I do not recall all the details of our conversation, but by the end, my colleague had convinced me to give the opportunity further consideration. At the time, I was serving as associate chair for graduate studies and, while I had toyed with the idea of becoming a department chair, I had no real interest in higher administration. I had previously served as vice provost for undergraduate education and as vice provost for academic diversity and was convinced that the real heroes were those who took on the leadership of departments—the most critical unit of our institutions.

On September 1, 2013, I began my appointment as the dean of GSOE becoming the first African American woman to serve as dean of engineering at a non-HBCU (Historically Black Colleges and Universities) institution. I was drawn to CCNY because of its mission as a public institution with a public purpose, founded in 1847, to provide access and opportunity. Seven years later, I joined Olin College of Engineering to serve as its second president. Founded in 1997 with an endowment from the Franklin W. Olin Foundation, Olin was created to transform engineering education by meeting the growing demand for engineers to have competencies in business and entrepreneurship and to better understand the social context of engineering. I was drawn to Olin because of its mission to "be an important and constant

---

G. A. Barabino (✉)
Olin College, Needham, MA, USA

J. P. Carpenter (ed.), *Women in Academia*, Women in Engineering and Science,
https://doi.org/10.1007/978-3-031-97938-5_10

contributor to the advancement of engineering education in America and throughout the world and, through its graduates, to do good for humankind."

Olin is one of a kind. From its founding, it was dedicated to project-based learning based on real-world applications, interdisciplinarity, as well as an integration of engineering, liberal arts, and entrepreneurship and gender parity in the student body. While Olin is a stand-alone institution dedicated to engineering and led by a president, in essence, due to its small size (400 students, 40 faculty) and lack of departments, it is more akin to a college or school of engineering within an institution. And further, its small size and emphasis on liberal arts and the social context of engineering make it akin to small liberal arts colleges.

Through my lens as an African American woman, after 7 years at CCNY leading GSOE and 5 years leading Olin, I believe that I am singularly positioned to provide a perspective on the engineering deanship that can shed deeper insights on the role and its implications, for women leaders in particular. In this chapter, I will focus on the path to the deanship, obstacles along the way and how to overcome them, what a deanship entails, leadership styles, and a vision for the future.

10.2 Scenarios on the Path to Deanship

I will begin with a set of hypothetical scenarios of lived experiences to illustrate and interrogate the landscape and its navigation for women in academic engineering. In reading the scenarios, I ask the reader to (re)imagine their own experiences and those of others they may be familiar with and consider the role of gender and other factors that influence experience and outcomes. Consider each scenario through the lenses of someone on the path and someone in the role of dean and ultimately consider how we can attract more women to leadership positions and ensure their success.

Scenario 1

Similarly talented and trained Professors Lesser and Moore started their careers at the same time in the same discipline, but at different types of institutions, Professor Lesser at a less-resourced teaching-intensive university and Professor Moore at a high-resourced research-intensive university. Their startup packages and first-year assignments were starkly different. Professor Lesser's modest startup package was split over 2 years and included summer salary and a reduced teaching load for 2 years. On the other hand, Professor Moore spent her first year carrying out a prearranged postdoc and enjoyed a startup package that included support for two graduate students, 2 years of summer salary and reduced teaching, an appointment in a research institute, and an assigned primary faculty mentor along with a mentoring team. Professor Moore's trajectory was advantaged by increased productivity in terms of publications and funding, while Professor Lesser's trajectory was stunted in terms of publication and funding metrics. Professor Moore eventually became chair of her department.

Scenario 2

As a junior faculty member in a department extoled for interdisciplinary research, Professor Peer was eager to share her ideas for grant proposals, to collaborate on research projects, and to co-advise students. She and Professor Steele had submitted a couple of proposals together, were co-advising a graduate student, and were both on the thesis committees for two students (one in her lab and the other in his). While Professor Peer was under the impression that the collaboration with Professor Steele was going well, that impression was shattered when she learned from one of her graduate students that a paper had been published on their collaborative work with Professor Steele as the senior author and that she and her student were left off from the authorship. Brazenly, Professor Steele bragged to her about a proposal that he had just received notice of funding for, and it was based on her ideas, yet she was not a co-PI. She did not even know that the proposal was submitted.

Scenario 3

Professor Redi had been at the rank of Associate Professor for 9 years and had committed significant effort in her service years prior and after promotion to associate leading the effort to develop and deliver a curriculum. She was ready to go up for promotion to full professor, and her tenure and promotion committee was lukewarm at best and her department chair, Professor Byess, thought that she should wait. He assured her that after one more year he would support her package. But then the next year, he told Professor Redi that another year would make her case stronger. Shortly thereafter, Professor Redi informed the department chair and the dean that she had accepted a position with another university.

Scenario 4

Professors Major and Minor, two senior women in the same department, both with active research programs (neither had previously served as department chair), happened to receive offers to become deans at different institutions in the same year. With the dean's approval, the department chair prepared a counteroffer for Professor Major who then turned it down as well as two subsequent counteroffers. Despite being similarly accomplished, Professor Minor, a woman of color, never received a counteroffer. Unsurprisingly, the arrangements for transition for the two professors were starkly different. Professor Major was afforded assistance and accommodations to make her transition as seamless as possible, while Professor Minor received no assistance and had to navigate unnecessary obstacles, including those that disadvantaged her graduate students.

Scenario 5

Dean Fairley was a newly appointed Dean of Engineering at Whiteman University and was strategizing on how to best use the faculty lines allotted to her as part of her hiring package, how to address long-standing tensions between centers and institutes and departments, and how to distribute centralized resources to meet unit demands. Her group of all male department chairs and mostly male center and institute directors made their intentions clear at the outset; each felt their unit had the greater need for a disproportionate share of the resources, and the directors purported a sense of autonomy that had allowed them to separate themselves from accountability around the collective good for the college of engineering. Dean Fairley was constantly challenged regarding her decisions related to resource allocations, including situations where individuals went around her to seek support directly from the provost and president.

10.3 The Path to Deanship

The predominant path to the deanship in America is academic and typically includes stints as a department chair or associate dean. In his 2018 survey article on paths to the deanship in academic engineering (Skinner, 2018), Richard Skinner noted that of the 186 deans in the survey, 18% were women, 74% were white, 29% were foreign born/educated, 39% held a department chair or head position immediately prior to their deanship (59% held a chair position at some point), 13% held an associate or assistant dean position, 63% joined a new institution, 41% joined a peer institution, and 44% joined an institution of lower ranking, and the most frequent institution where engineering deans earned their doctorate was MIT (13 of 186).

Ascension through the academic ranks serves as a major gatekeeper as to who may eventually become a dean. Within each of the above scenarios, it is apparent that there are myriad factors that influence trajectories and there are differential experiences dependent upon institutional types, access to resources and opportunities, collaboration, and who is in charge, among other variables. Women like Professors Lesser, Peer, and Redi (Scenarios 1–3) are disadvantaged by exclusionary practices and behaviors, limitations related to access to resources and opportunities, and detrimental decisions made by those in charge. Gender differences are pervasive, as shown by Araujo et al. (2017); men are more likely to collaborate with men, and women are more egalitarian in scientific collaborations. Even at the senior level, biases and obstacles persist, particularly for women of color as experienced by Professor Minor (Scenario 4), for example.

The role of doctoral prestige in faculty hiring is pointed out by Aaron Clauset, K. Hunter Wapman, and their colleagues who note that the prestige of an individual's doctoral institution is a predictor of how well one will be placed in the hiring hierarchy (Clauset et al., 2015; Wapman et al., 2022) and landing in a higher prestige institution leads to increased productivity (imagine Scenario 1). Further complicating matters for underrepresented minority faculty, they receive more negative tenure and promotion votes and are less likely to receive unanimous positive votes, the discrepancy being attributed to a double standard (Masters-Waage et al., 2024). And what is the impact on retention? In a comprehensive study of faculty retention across the United States, Spoon and her colleagues revealed that women are more likely to leave and less likely to be promoted than men; departmental environments influence retention; prestige mitigates attrition; and gender, discipline, career stage, and institution account for differential experiences among faculty (Spoon et al.,

2023). With respect to the receipt of counteroffers, White-Lewis et al. (2024) demonstrated that white faculty were 64% more likely to receive counteroffers than faculty of color, men were 80% more likely to receive counteroffers than women, and faculty at less research-intensive universities were 80% less likely to receive counteroffers than their counterparts at research-intensive universities.

Specific to engineering, Ermagun and Erinne describe a "virtuous cycle" of favorability for higher-ranked institutions where the elite get more and a "vicious cycle" for lower-ranked institutions where they become entrapped in a state of disadvantage. Evidence for this exclusionary system is the skewed hiring patterns related to rankings. Roughly 70% or more of engineering faculty hired at top 20 institutions received their PhDs from top 20 institutions (Ermagun & Erinne, 2022). Employing social capital theory as a framework for their examination of trajectories for Black engineering and computing faculty to university leadership, McGee and colleagues found the predominant path to be academic and characterized by intentionality regarding a desire to increase representation and serve as a role model (McGee et al., 2022).

Career mobility and trajectories for women are impacted by race/ethnicity, gender, prestige hierarchies, and social capital. Ascension to a deanship can be in reach or out of reach depending upon a multitude of factors, as illustrated in the scenarios provided and backed up by extant literature.

10.4 The Deanship

Deans are ideally situated within the leadership of an institution to enact mission and affect change. In the middle between upper administration (presidents and provosts) and faculty, and serving as the public face for their unit, deans interact with all facets of the institution, its internal constituents and external groups. They are critical for developing and carrying out the institutional vision and mission; overseeing the attraction and retention of students, faculty, and staff; setting and meeting academic standards and goals; setting policies and ensuring adherence to them; and ensuring financial sustainability.

As stewards of the profession, deans of engineering are acutely aware of linkages between the well-being and security of our nation, engineering-driven innovation, and broadening participation in the field. Thus, they are uniquely positioned to provide leadership in the development of a diverse cadre of engineers who are innovators, leaders, and problem-solvers. They are also well positioned to play a critical role in encouraging and developing a diverse cadre of faculty members that are poised to become deans in the future by providing leadership and other opportunities for professional development.

Women engineering deans, while still outnumbered by men, continue to make strides as our numbers increase and their influence grows. Recent data collected by the Society of Women Engineers noted that as of 2021, women held 20% of engineering dean positions.

There are myriad reasons why women lag men in acquiring dean positions, some of which are related to the differential experiences of women, particularly women of color, in academic engineering along schooling and academic career trajectories. In addition to factors that impact pathways to the deanship (as discussed in the previous section), Henningsen et al. (2021) answered the question: Where are the women deans? Their study points to two possible explanations for the underrepresentation of women holding dean positions: gender bias and self-selection bias (Henningsen et al., 2021). For many women, the role of dean just is not that appealing. For women of color, and Black women in particular, persistent barriers rooted in structural racism result in limited access to networks, advocates, and opportunities (McGee et al., 2021).

A better understanding of the experiences of women deans can be gained from the profiles of women deans carried out by the Society of Women Engineers (SWE) over two decades starting with Eleanor Baum, PhD, the first woman dean of engineering in the country in 1984. Building on these profiles, in her 2010 article, "Perspectives on Leadership from Female Engineering Deans," Peggy Layne discussed what deans do and the paths to the deanship based on interviews with 21 women deans (Layne 2010). The interviewees noted the responsibilities of the dean ranging from representing and advocating for their college/school to faculty recruitment and retention to personnel and budget management. Their paths to the deanship were varied, and among these early women deans, preparation took various forms and very often did not involve a previously held department chair position.

Layne's 2021 article, "Women Engineering Leaders in Academe 2021," takes a retrospective look at profiles of women engineering deans over the years including a look at "where are they now?" and experiences along the way (Layne, 2021). While paths varied, women engineering deans had several things in common: desire and ability to work well with people, collaborative/nurturing leadership style, willingness to take on new career opportunities, and willingness to be themselves.

10.5 Leadership

Who leads determines who will lead. Leaders set the example and make career-impacting decisions that can either open or close doors. At a minimum, those serving in dean positions should provide healthy scrutiny over tenure and promotion decisions advanced by departments and should understand cases that are not advanced (the same level of scrutiny should be applied for hiring). They should examine criteria like time in grade, course loads, service loads, and access to graduate students to look for and uncover patterns that can be discriminatory leading to differential outcomes.

In the case of Professor Redi (Scenario 3), Department Chair Professor Byess lost a contributor whose loss could have lasting implications for the department. Most often, but not always, the chair is guided by the recommendation of a departmental level tenure and promotion committee whose membership may or may not include a member with sufficient understanding of the candidate's area of expertise or the biases that women candidates face. Further, data (both internal and external records) that could be enlightening is absent.

At the dean's level, a step removed from and above the department level, broader comparisons can be made, and there is yet another opportunity to check for biases intentional or not that may have been present in the evaluation.

I recall a specific instance while serving on a tenure and promotion committee when two cases were presented for candidates with extensive interdisciplinary work. When the male candidate's case was presented, committee members credited him with being a central contributor to the interdisciplinary group effort. To the contrary, for the female candidate, the committee members questioned the significance of her contributions and implied that she was being carried by the group of collaborators, despite her strong research funding record as a principal investigator (much stronger than that of the male candidate).

As a dean, I was always cognizant that these types of biases creep in and often go unchecked; I knew that there was not always someone on the tenure and promotion committee who would play the role that I played as a committee member in speaking up.

As described in Scenario 5, women deans of engineering are often in majority-male environments and faced with challenges from mostly male direct reports who resist their leadership. In their Harvard Business Review article, Eagly and Carli address resistance to women's leadership related to differential perceptions and associations regarding leadership based on gender, for example, women being seen as more nurturing compared with men (Eagly & Carli, 2007). They and others believe that there is a distinct female leadership style. Women tend to exhibit characteristics of transformational, communal/collaborative, and nurturing leadership (Eagly & Carli, 2007; Eagly et al., 1992). Operating at the intersection of race and gender, women of color cannot escape the double standards of the glass cliff effect or intersectional invisibility and its associated disadvantages, often overcome in part by leaning into authentic leadership (Smith et al., 2018).

Beyond their leadership styles which tend to be collaborative and transformational, women leaders play critical roles as mentors and role models; as vision and mission setters; and as developers of the next generation of leaders.

10.6 Vision for the Future

Deans are uniquely positioned to affect change and promote institutional success as midlevel administrators; are well positioned to join the highest ranks of university leadership; and are critical to the training of future generations of faculty and practitioners. Women bring distinctive perspectives and approaches to the deanship, yet they remain starkly underrepresented. Preparing more women to embark on the path to becoming a dean and ensuring that the path is not littered with obstacles are a higher education imperative. Success in this arena requires a broader aperture and criteria for entry and selection at each rung along the career path; a reevaluation and redefinition of what are viewed as requisite skills and how those skills are attained; encouragement for some women to move past their first inclination to opt out of leadership opportunities; collection and disaggregation of contextualized data on women; and intentionality along with accountability in ensuring career success.

I have lived experience in academic engineering as a dean and on either side of the deanship as a Black woman that informs my vision for the future. I have had to navigate rough terrain for myself and for others. And like other Black women (and other women operating at the nexus of race and gender), I have found that the experience I gained in overcoming barriers, resisting inequitable systems, and navigating inhospitable environments helped me develop what has been referred to in the business literature as resistance and navigational capital (Jackson & Rajai, 2021). My presence and the presence of other women and women of color have helped in the reimagination of what a deanship is and what a dean looks like.

The future I envision will reflect institutional and systemic change that allows for the sustenance of a culture and environment in which opportunities match talent; recognition and rewards that match accomplishments and those responsible for them; policies and practices that are fair and just (not privileging some while disadvantaging others); disparities in access and advancement that are eliminated; prestige that does not dictate opportunities; and no hierarchical structures that limit opportunities for women and others. In the future I envision, we will no longer need to ask the question: "Where are the women deans?"

References

Araujo, E. B., Araujo, N. A. M., Moreira, A. A., Hermann, H. J., & Andrade, J. S. (2017). Gender differences in scientific collaborations: Women are more egalitarian than men. *PLoS One, 12*(5), e017691.

Clauset, A., Arbesman, S., & Larremore, D. B. (2015). Systematic inequality and hierarchy in faculty hiring networks. *Science Advances, 1*(1), e1400005.

Eagly, A. H., & Carli, L. L. (2007). Women and the labyrinth of leadership. *Harvard Business Review, 85*(9), 62–71.

Eagly, A. H., Makhijani, M. G., & Klonsky, B. G. (1992). Gender and the evaluation of leaders: A meta-analysis. *Psychological Bulletin, 111*(1), 3–22.

Ermagun, A., & Erinne, J. (2022 Dec). A systematic exclusion induced by institutional ranking in engineering faculty hiring: Introducing a cycle of winners and losers. *PLoS One, 17*, e0275861. https://doi.org/10.1371/journal.pone.0275861

Henningsen, L., Eagly, A. H., & Jonas, K. (2021). Where are the women deans? The importance of gender bias and self-selection processes for the deanship ambition of female and male professors. *Journal of Applied Social Psychology, 52*, 602–622.

Jackson, M., & Rajai, P. (2021). Does your definition of leadership exclude women of color? *Harvard Business Review*.

Layne, P. (2010). Perspectives on leadership from female engineering deans. *Leadership and Management in Engineering, 10*(4), 185–190.

Layne, P. (2021 Fall). Women engineering leaders in academe 2021. SWE Magazine.

Masters-Waage, T., Spitzmueller, C., Edema-Sillo, E., St. Aubin, A., Penn-Marschall, M., Henderson, E., Lindner, P., Werner, C., Rizzuto, T., & Madera, J. (2024). Underrepresented minority faculty in the USA face a double standard in promotion and tenure decisions. *Nature Human Behaviour, 8*, 2107–2118.

McGee, E. O., Jett, C. C., & White, T. T. (2022). Factors contributing to black engineering and computing faculty's pathways toward university administration and leadership. *Journal of Diversity in Higher Education, 15*(5), 643.

McGee, E. O., Main, J. B., Miles, M. L., & Cox, M. F. (2021). Intersectional approach to investigating persistence among women of color tenure-track engineering faculty. *Journal of Women and Minorities in Science and Engineering, 27*(1), 57–84.

Skinner, R. A. (2018). Paths to the deanship in American academic engineering. *Bridges, 48*(4), 53–58.

Smith, A. N., Watkins, M. B., Ladge, J. J., & Carlton, P. (2018). Interviews with 59 black female executives explore intersectional invisibility and strategies to overcome it. *Harvard Business Review*.

Spoon, K., LaBerge, N., Wapman, K. H., Zhang, S., Morgan, A. C., Galesic, M., Fosdick, B. K., Larremore, D. B., & Clauset, A. (2023). Gender and retention patterns among U.S. faculty. *Science Advances, 9*(42), eadi2205.

Wapman, K. H., Zhaong, S., Clauset, A., & Larremore, D. B. (2022). Quantifying hierarchy and dynamics in US faculty hiring and retention. *Nature, 610*, 120–127.

White-Lewis, D. K., O'Meara, K., Mathews, K., & Havey, N. (2024). Counteroffers for faculty at research universities: Who gets them, who doesn't, and what factors produce them? *Higher Education, 89*, 535–552.

Chapter 11
Navigating Engineering and Academia: The Journeys of Four Intersectional Women Leaders

Kelly J. Cross, Idalis Villanueva Alarcón, Olga Pierrakos, and Denise R. Simmons

11.1 Introduction

11.1.1 Motivation

From roots in the rural south to cross-country moves to the United States, one's upbringing shapes the lens by which their personal, educational, and professional lives unfold and how others interact with an individual. Rooted in complex ancestries and histories, gender-, racially, and/or ethnically diverse academic leaders are continually confronted with new challenges that place them at the crux of multiple epistemic injustices and other forms of prejudices around their capabilities and performance. Epistemic injustices are severely understudied in engineering, in part due to limited understanding of how societal, political, and historical intersectionalities contribute to a person's self-concept of their leadership potential and ability. This chapter serves as an exposition of our reflections on our paths to academic leadership to uncover potential areas that warrant additional exploration. The authors hope that the reflections provided will bring hope and inspiration to those who are also victimized by these forms of epistemic injustices. The authors also recognize that leadership shows itself formally and informally throughout all aspects of their lives.

K. J. Cross (✉)
Georgia Tech and Emory University, Atlanta, GA, USA
e-mail: kelly.cross@bme.gatech.edu

I. V. Alarcón · D. R. Simmons
University of Florida, Gainesville, FL, USA
e-mail: i.villanueva@ufl.edu; simmons@eng.ufl.edu

O. Pierrakos
Wake Forest Engineering the National Science Foundation, Alexandria, VA, USA
e-mail: pierrao@wfu.edu

J. P. Carpenter (ed.), *Women in Academia*, Women in Engineering and Science,
https://doi.org/10.1007/978-3-031-97938-5_11

11.2 Framing the Problem

In STEM higher education, reports have pointed to the ongoing and limited representation, promotion, wage fairness, and retention of women in engineering, particularly among those who have intersecting and diverse racial and ethnic identities. According to the 2021 Women, Minorities, and Persons Disabilities in Science and Engineering report (National Science Foundation [NSF] 2021), women in underrepresented racial and ethnic groups comprise less than 2% of women faculty in tenure or tenure-track positions. Specifically, 1.4% of STEM faculty who are tenured are Black women (Singh, 2020) while 1.3% are Latiné/x women (Singh, 2020) compared to White women faculty (Singh, 2020; Society of Women Engineers [SWE] 2023). Pronounced wage gaps between women, women of color, and men in the engineering faculty lines have also been found where, on average, women faculty earn less than men faculty members, and Black and Latina women faculty are paid significantly less than White and Asian American women (McGee et al., 2024).

Disparities among women of color in engineering are measurably more pronounced when using artificial intelligence (AI) research methods, as recently shown in a Nature publication (Pogrebna et al., 2024). In Pogrebna et al.'s (2024) study, the authors quantifiably demonstrated that Black women academic leaders representing diverse racial minorities tend to engage in higher risk-taking behaviors to make systemic or structural changes; often, said changes are not implemented. As Pogrebna et al. (2024) suggested, there were multidimensional challenges stemming from a "legacy of systemic discrimination" (p. 14) that limited how these academic leaders perceived the world and their placement in that world.

Furthermore, recent work from McGee et al. (2024) has begun to advocate for a more qualitative approach to uncover the layers that contribute to these and other types of systemic oppression for academic women leaders in engineering. In their study (McGee et al., 2024), the authors used qualitative approaches to explore the experiences of women academic leaders in engineering and found that wage disparities were more pronounced for Black and Latina women of color. McGee et al. (2024) were motivated to conduct the study qualitatively due to multiple factors including but not limited to the rising cases of academic bullying, departures of women of color faculty (either by silent quitting, unsubstantiated firing, or even loss of life), and recent decisions to reaggregate national data concerning the career trajectories of academic women in STEM (e.g., National Science Foundation [NSF] 2023).

Whether quantitative, qualitative, or both, there is a need to uncover the real experiences of women academic leaders in engineering with diverse intersectional identities. What is unclear at this point are the experiences and reasons *why women in engineering who are diverse* racially, gender-wise, and ethnically stay or the life experiences that motivated them to persist. It is known that systemic discrimination, from an intersectionality lens, is rooted in historical, societal, legal, and political norms that disproportionately and negatively impact minoritized women in society (Crenshaw, 1990). But *how did women in engineering with diverse identities, at the intersection of their personal and professional experiences, navigate and overcome the epistemic injustices they have faced?* This chapter aims to help fill that gap.

11.3 Theoretical Frameworks

11.3.1 Epistemic Injustice

Epistemic injustice is defined as the phenomenon in which an individual receives unjust treatment based on negative perceptions of their *capabilities as a knower*. The theory of epistemic injustice originates in feminist philosophy rooted in postmodernism (Fricker 2000, 2007) and manifests in two types: testimonial and hermeneutical. Both forms are distinct in definition but intertwined within social interactions.

Testimonial injustice is a type of epistemic injustice in which a "prejudice causes a hearer to give a deflated level of credibility to a speaker's word" (Fricker 2007, p. 1). The process of testimonial injustice takes place between two individuals—the hearer receiving information and the speaker giving information—within a testimonial or informational exchange. For a productive testimonial exchange to take place, the hearer must make a credibility judgment, in which they believe some level of truth and trustworthiness from the speaker to determine the information as true and valuable (Lackey 2006, 2008; McKinnon 2016). Sources the hearer can use to make this credibility judgment can come from "information about the speaker themself [including visible identities ...], information about how the speaker said what they did [...], background knowledge on the content of the assertion [...], and information about the structural features of the assertion" (McKinnon 2016, p. 437). Each of these sources offers opportunities for implicit biases, or unconscious ingrained thought patterns, prejudices, or stereotypes (Judson et al. 2017), to be used in credibility judgments. If the hearer judges the speaker as having credibility over what the speaker legitimately has, the hearer commits an epistemic injustice to the speaker by incorrectly judging their credibility in the context of the testimonial exchange. The accumulation of testimonial injustices can skew the credibility of the speaker to reflect systemic issues of gender, race, sexuality, or socio-economic status (SES) inequality within a culture.

Hermeneutical injustice is a systemic gap in resources that unfairly disadvantages a group in making sense or meaning of their personal experiences (Fricker 2007; McKinnon 2016). A current example of hermeneutical injustice is the push for downplaying or eliminating discussions about slavery and racism from educational curricula in southern US states, which leaves students to make meaning of the trauma from these experiences. Though this current affair is frontline news today, the act of withholding knowledge and knowledge resources from the Black and Indigenous communities in the United States has occurred since the colonization of the Americas (Callimachi 2021; Williams 2009; Woolford 2015). Because hermeneutical injustice is systemic, these gaps in knowledge and resources form the context for social interactions that can manifest into testimonial injustice. For example, when situating research within engineering learning environments, the underlying knowledge gaps that stem from white supremacy, patriarchy, colonization, and other systemic ideologies define the dynamics of social power instances of testimonial injustice. Therefore, though testimonial and hermeneutical injustices are analytically distinct, they interact and influence each other because neither of them can be separated from the context in which they are situated.

The engineering education field can benefit from the scholarship on epistemic injustice, and the theory should be operationalized in engineering contexts that lack the broad representation of gender, race, and ethnicity consistent with our society. Though a relatively new theory to engineering education research (EER), in the traditions of philosophy, psychology, and education, epistemic injustice has been used to call out the exclusionary or marginalizing practices within cultural or institutional systems (Bhakuni and Abimbola 2021; Dotson 2014; Settles et al. 2021). Apart from epistemic injustice research, STEM scholarship broadly on women's underrepresentation in engineering is under-discussed (Beddoes and Borrego 2011). Women in engineering have experienced prejudice in many forms, in part due to their gender identity and engineering's exclusionary culture toward women with diverse identities (Beddoes and Borrego 2011; McLoughlin 2005; Ong et al. 2020).

11.3.2 Intersectionality

Dr. Kimberlé Crenshaw's seminal work (Crenshaw, 1990) described the social injustices that African American women were experiencing due to the intersection of race and gender in different legal systems. Almost 37 years after introducing the term *intersectionality*, Dr. Crenshaw advocated for a state of urgency for scholars to study intersectionality properly as she posited that many times, the term is improperly misunderstood or misused to signify additive categories to a person's identity (e.g., race + gender) rather than uncovering the systems that oppress people with intersecting identities (Crenshaw, 2017).

Used as a way to understand the human experience in society (Collins & Bilge, 2016), intersectionality provides a definition and framework that helps understand how the presence of politics, culture, class, wealth, privilege, power, citizenship, birthright, and other *-isms* due to race, gender, class, heterosexuality, transphobia, xenophobia, and ability, among others, compounds to the inequalities and injustices society faces today (Crenshaw 1989, 1990, 2017). We would be remiss to not point out that in engineering education, ***intersectionality is often misused and interpreted as the addition or intersection of many categorized identities, instead of used in its original form and intent*** (Bowleg, 2008; Collins, 2015). Engineering education studies, particularly for women in engineering with diverse identities, must consider the roots and intentions of Crenshaw's intersectionality framework (1989, 1990, 1991, 2017) to properly situate studies to uncover systemic issues rather than applying deficit, additive classifications that further the epistemic injustices experienced by marginalized groups.

11.3.3 Self-Concept Theory

Deriving from social psychology, self-concept involves a cognitive approach to perceiving oneself as a subject of interest in exploring oneself as a phenomenon (Dember, 1974; Manis, 1977). A person's internalized interests may vary to the degree to which said individuals situate life events (selected by the individual) as being central to their self-concept and identity (Rubin, 2020). Individuals' processes involved in autobiographical memory require that numerous personal and professional identities become salient, which allows them to respond quickly and flexibly to changing circumstances and different societal roles (Campbell et al., 2003). As an example, an engineering female student of color in a white male-dominated classroom environment will be more aware of their gender and racial identities than the engineering student identity. In contrast, an engineering female student of color in a racially focused sorority will be more aware of her engineering student identity. Social contexts cannot be ignored in understanding the salience of one's multiple identities.

Similarly, context and salience apply to leadership identity as well. Autobiographical memories may also be situated at an earlier or future place and time where they may have been exploring themselves or being themselves (Rubin et al., 2019). Thus, self-concept theory postulates that individuals may position their stories and narratives to those events that may have influenced specific roles in society (Campbell et al., 2003).

In the context of epistemic justice and intersectionality, self-concept allows the person to position themselves amid complex, oppressive systems and contend with the identities and roles that may become salient or grounded in several situations.

11.4 Research Design

This chapter is premised on the principles of qualitative research designs, more specifically around a collaborative autoethnography process in the context of engineering (e.g., Cross et al., 2022; Secules et al., 2021; Sochacka et al., 2016). Autoethnography includes a systematic approach to analyzing a person's own experience, as recalled autobiographically by the individual, and connecting it to a larger meaning framework (e.g., culture, history, politics, society) (Costello et al., 2016; Ellis et al., 2011). While it includes elements of autobiography (personal data of an experience), its situating of sociocultural realities (as those typically found in ethnography) allows scholars to identify pivotal transitions and timepoints important in an individual's life trajectory (Ellis et al., 2011).

The end goal of autoethnography is to create a counter-narrative on the ways that individuals resist oppressive structures and enact social change (Solórzano & Yosso, 2001). When used collectively, collaborative autoethnography, the combination of voices and experiences, serves a given phenomenon (Chang et al., 2014) and a person's self-concept connecting to the epistemic injustices faced at the intersectionality of their personal and professional contexts.

11.4.1 Positionality

All four authors come with diverse and rich experiences that intertwine faith, family, culture, language, economic backgrounds, history of oppression, and gendered and racialized ancestries across diverse professional and academic social contexts.

All recognize the privilege that they have uniquely had, and all collectively recognize and value the roots and upbringing that shaped their academic leadership.

All authors have differing levels of educational experiences, both in their originating country or culture and in the United States. All recognize that their identities are non-monolithic (Revelo et al., 2017), are based on practices and dimensions rooted in reflexivity and context, and are used to minimize bias and improve research quality (Secules et al., 2021; Martin et al., 2022). While for this work, we will derive suggestions on areas that warrant more exploration in engineering education, the authors decided to maintain their stories authentic to their identities and stylistic preferences to help the reader situate the pivot points that led to transformational changes and decisions despite the epistemic injustices experienced. We also want to point out that each author decided to frame the positionality that was most salient to their self-concept and experience. As such, formatting may be different for each author because we are all non-monolithic (Revelo et al., 2017).

Dr. Kelly J. Cross is Assistant Professor in the Wallace H. Coulter Department of Biomedical Engineering (BME) at Georgia Tech and Emory University. Dr. Cross self-identifies as a data-informed, transformational mission-focused culturally responsive practitioner, researcher, and educational leader. Being raised in a very religious home where her father stressed loving God's people and her mother emphasized living a "Godly" righteous life, Cross's parents laid the foundation for her core sense of self rooted in spirituality and provided many life lessons early in her childhood. This foundation has played a significant role in Dr. Cross' identity development throughout her academic and professional career.

Dr. Idalis Villanueva Alarcón (her last name includes the latter two surnames) is a tenured Associate Professor and Associate Chair in the Department of Engineering Education at the University of Florida. She has a Ph.D. in Chemical and Biological Engineering and a postdoctoral degree from the National Institutes of Health and led an academic career first as a lecturer and then as a tenured faculty in engineering education. Born and raised in Puerto Rico, she was always in between cultural and linguistic identities that created in her a need to code-switch and interpret hidden curriculum surrounding her environments. Rooted in her faith in Jesus of Nazareth and her family, she carries lessons and principles to abide by when navigating her academic leadership roles and responsibilities.

Dr. Olga Pierrakos is currently a Program Director at the National Science Foundation (NSF) in the Directorate of STEM Education as well as Founding Professor of Wake Forest Engineering. Prior to her role at NSF, Olga served as the Founding Chair of Wake Forest Engineering (2017–2022) having the responsibilities of what would be an Engineering Dean at most other universities. Olga and her team set out to reimagine and redefine engineering education with a vision to educate the whole engineer with a commitment to humanity and human flourishing. The successes of her leadership positioned Wake Forest Engineering to become that highest ranked (US News Report), most diverse (gender and racial/ethnic diversity of students and faculty), and most innovative (curriculum, pedagogy, research) academic unit at Wake Forest University. Olga joined Wake Forest University after a 2-year Program Director rotation at the NSF and having served as founding faculty at James Madison University. Both Wake Forest University and James Madison University being predominantly white institutions (PWIs), Olga notes many differences in her experiences at these two institutions in contrast to her leadership roles navigating the National Science Foundation, which represents a vast range of racial, ethnic, and gender diversity. Olga's previous experiences as a Greek immigrant (coming to the United States at the age of 10 with no English), first-generation college student, first-generation engineer and Ph.D., and mother of four have shaped her unique lens on academic leadership.

Dr. Denise R. Simmons is an Associate Dean for Workforce Development at the Herbert Wertheim College of Engineering and a tenured Professor in the Department of Civil and Coastal Engineering in the Engineering School of Sustainable Infrastructure and Environment (ESSIE) at the University of Florida. Denise is driven and inspired by what an organization can achieve by attending to people's needs. This passion was influenced by her South Carolina upbringing, the women in her family, her work experiences, and her faith.

11.4.2 Methods

The four authors of this study served as researcher participants (Crenshaw, 2017) and discussed the scope of this chapter from the lens of epistemic injustice, intersectionality, self-concept theory, and academic leadership in engineering. All agreed to respond to a set of questions drafted by the first and second authors and corroborated by the remaining authors. From the discussions, questions were edited and finalized into the following:

1. Tell us a little bit about your upbringing and what you consider makes you intersectional.
   (a) Please provide the context of the historical, political, or legal realities of that upbringing, if applicable.
2. Who am I?
3. Explain a bit more about the academic leadership position(s) you have or have had. Explain to each one how you have brought your unique personal identity into these positions. Discuss any critical incidents that led to your leadership style.
4. How have your knowledge and expertise been valued in academic leadership spaces? How has it not been valued in your academic leadership spaces?
5. In each of your academic leadership position(s), how have you maintained your life goals, relationships, and well-being despite oppressive climates or situations? Please provide contexts into these forms of oppression and provide strategies you have used to maintain yourself intact in the process.
7. How do you view your academic leadership(s) has served to inspire others?
8. What recommendations or strategies do you have for other intersectional identities especially to those who have similar shared identities as you to "become academic leaders in engineering"?
9. Describe any instances where your personal and professional identities were at odds with each other during your academic leadership roles.
10. Add anything else you wish to share.

The authors had approximately a week to respond to the questions that were included in a Word document embedded within an encrypted Teams folder. Each participant generated a copy of the questions and provided their responses in written form. The first author collected the written responses and performed an initial round of inductive and axial coding (Saldaña 2016) to generate a draft codebook, and this was verified by the second author. The additional authors verified, member-checked, and agreed upon the codebook. All authors met to discuss and identify common themes and points of intersections of the autoethnographies. Themes were discussed from the standpoint of epistemic injustice and its commonalities and deviations, and discussions took place until consensus was achieved. A list of strategies for academic leadership was generated. To ensure the trustworthiness and quality of the data, we followed a similar procedure established by Cross et al. (2022) where they created a set of standards adapted from Chang (2016) but tailored to the experiences of individuals with various intersecting identities in engineering.

11.5 Findings

The findings presented consist of the commonalities and differences of each participant in this narrative. While we recognize that we all have intersecting identities, our walks of life and profession are different. The themes discussed present the pivotal points of our stories.

Theme 1: Epistemic Injustices Stem from Self-Concept of Their Intersectionalities

All authors commonly described having a unique upbringing compared to the traditional familial backgrounds of majority groups in the United States (e.g., White, middle class, American dream). While many of the authors recognized that privilege existed in their upbringing compared to similar counterparts in their racial or ethnic communities, they also recognized the inequities that surrounded their transitions from childhood to adulthood. By presenting this theme, the authors posited that self-concept situates epistemic injustices that are rooted in their intersectional systems and backgrounds. In other words, turbulent histories, cultures, and political and legal systems within their surrounding environments created self-concepts of how they should behave and act in society, leading to epistemic injustice as they navigated different systems later in life.

> *My upbringing as an African American girl born in late 1960s Charleston, South Carolina was profoundly influenced by the turbulent historical, political, and legal realities of the time. I navigated the* ***challenges of racial segregation, economic disparities, and the enduring legacy of the Civil Rights Movement*** *while drawing strength from my* ***cultural heritage, family, and community bonds.*** *Growing up in Charleston meant confronting* ***the harsh legacy of segregation*** *as the city was a major center for selling and enslaving Black people. It took me a long time before I realized that the Market,* ***a celebrated tourist attraction in downtown Charleston, was formerly known as the Slave Market.***—*Simmons*

> *While I was raised in front of a coastal area across the street to a beach in Puerto Rico [...] the surrounding areas to* ***our house were full of poverty and drug usage*** *... in the* ***peak of the AIDS crisis*** *and the rise of drug and illicit substance and* ***psychoactive drug abuse***

> ***(e.g., cocaine, heroin) in the United States, extending to its territories (i.e., Puerto Rico)*** *[...] That meant that* ***from an early age, I was taught to be careful about my surroundings and people.****—Villanueva Alarcón*

> *I consider myself intersectional based on my multiple integrated identities as a* ***Black female queer-identified engineering professor with strong religious beliefs*** *[...] my parents* ***emphasized academic excellence although I didn't know any engineers growing up and higher education was not discussed.*** *Finally, I was part of the* ***desegregation program in the Midwest town*** *I grew up in and that educational experience* ***sensitized me to race early in my life.*** *As a result, a great deal of my life* ***has been exploring how I define myself or self-determination and how are my values and behaviors consistent with those identities.****—Cross*

> *As a naturalized US citizen, at the age of 10, I came from a homogeneous culture and society (Greece) with* ***strong national and religious identities to a more diverse culture and society*** *in the United States.* ***Navigating these distinct cultural identities*** *was a* ***daily reminder of not belonging and being an outsider*** *during a critical time during adolescence (middle school and high school). I was a* ***cultural outlier in all academic environments (K-12 and college).****—Pierrakos*

The experiences described by the authors highlight inequities or challenges of their upbringing that taught them early on to stratify their identities based on demographics or social constructions of designations of preference based on physical features or social conditions. All were judged or had a negative experience based on factors not related to academic ability or intellect. This is an example of epistemic injustice (EI) because the field of engineering operates under a false notion of meritocracy, where equal effort results in equal reward. All were discounted in some way based on their upbringing or view of them being an outsider.

Theme 2: Epistemic Injustices Create a Double Consciousness: Neither Here nor There

Many of the authors explained being in between two worlds: majority and minority, leading to a double consciousness (Du Bois, 1897, 1903). Some of the author's stories were tied to nationalities and cultures, whereas other expositions shed light into the systemically and historically generated realities that many were forced to live through. The liminalities that each of the authors experienced created a fragmentation or shock into their systems (Mejia et al. 2022) that revealed to them an important reality that prepared them to face life and leadership in the years to come.

> *As an* ***immigrant in the United States, I embraced two cultures that have defined me and have enabled me to bridge the best of each culture.*** *My parents divided across two countries* ***only had high school degrees but succeeded through hard work.*** *Raising four kids (myself being the second oldest), we all* ***navigated two cultures, two lands, two countries. Success was not an option; it was a necessity of survival.****—Pierrakos*

> *In addition to my parents,* ***my K-12 academic experiences shaped me as a scholar, researcher, leader, and educator.*** *Participating in* ***state state-sanctioned desegregation program in a racist Midwest town sensitized me to my race and racial disparities at an early age.*** *Growing up in the church and the middle class,* ***I didn't initially connect race, education, and socioeconomic status (SES).*** *During my doctoral program in engineering education is* ***when I learned that SES is still the most consistent predictor of academic performance and achievement,*** *which ultimately can impact access. This is important to note as we know the* ***US history of red-lining with housing and restricting access to education based on race.*** *I was* ***one of the few Black students bused to a PWI in an affluent area, that was able to successfully navigate the balance of academic performance and the cultural wars between the races.*** *There was* ***very little guidance to how we as students***

> ***could embrace or take advantage of the racial and cultural differences created by the bussing program. This was also my first time witnessing how systems of oppression work together.*** *I was able to observe* ***up close and personal how race and SES can be used to harm students from previously excluded ethnicities and racial groups****. As a result, I wanted to explore race in the context of the field of engineering and particularly within the training or engineering education. My awareness of inequity in education based on race is a core component of my research agenda.—Cross*

> *Being raised in the island of Puerto Rico, I quickly learned that* ***while we are a part of the U.S., we do not fully belong****. In between* ***Latin American cultures, Spaniard colonized languages, products of West African slavery, genocide of our native Caribbean Indigenous roots, and American commonwealths****, I am* ***constantly in liminality*** *(neither here nor there). I have always been living in an* ***in-between space****.—Villanueva Alarcón*

Based on the intersectionality surrounding their identities, each author discussed managing the liminality or the double consciousness as defined by Du Bois (1897, 1903). These dichotomies created lenses by which they perceive themselves and their positioning in the world (Rubin, 2020); this self-concept guided the type of academic leader they would become.

Theme 3: Self-Concept of Academic Leadership Serves to Mitigate Epistemic Injustices

Many of the authors described their intersecting identities and the intersectionality of their backgrounds being their superpower. These self-concepts became a major impetus to navigate their engineering landscapes connected to their trajectories. Having a strong self-concept resulted in authenticity and boldness in the way they approach leadership.

> *I have founded two brand new engineering programs and the intersectionality of being an* ***immigrant, a woman, an engineer, a humanist, a leader, an advocate, a mother has shaped by every fiber*** *[…] Who I am is most certainly a* ***part of every role I have undertaken****. As Founding Chair of Wake Forest Engineering, I* ***showed courage and responsibility to challenge traditions, norms, structures, and practices****. This was not an effort to challenge for the sake of challenging, but to* ***responsibly question the traditional ways we have operated in academia so that we can reimagine a new vision, a new engineering education****. I saw for far too long how silence, complacency, bias, and exclusion overtake higher education. During my negotiations to accept the position, I* ***fully disclosed that I had no interest to uphold the status quo of engineering education. Copy/pasting what has existed was not my mission*** *and I made that very clear upfront. My leadership style is* ***inclusive, collaborative, empowering, and responsible.*** *The successes of the team and department I led serve as evidence of the innovation, inclusion, and integrity to meet the needs of our diverse students. I knew that I would need to model as a leader what it is I expected of the kinds of engineers and professionals we were trying to produce.—Pierrakos*

> *My personal intersectional identities were critical to the* ***process as I had to select collaborators that were strong enough to withstand the default resistance from STEM fields about the topic****, but also with the* ***emotional intelligence and skill to create a safe space for myself and fellow engineers to explore a topic that we are not accustomed to broaching****. So, selecting the [research]team was critical and although I was the "leader", I was very clear that* ***my colleagues had technical knowledge about privilege and power that I did not have****. So,* ***their expertise allowed me to be vulnerable and interrogate what I didn't know or understand*** *(e.g. I had to embrace the role of the learner even as a leader). My privilege team provided me with an opportunity to be a servant leader as my parents would say.* ***As a servant leader, my colleagues showed me grace to train and assist in my learning and we as team provided space for engineers to reflect on how power, privilege, and collusion plays out in our daily professional activities, including teaching or mentoring.****—Cross*

Amidst the challenges of navigating corporate America's ingrained biases, my journey ***from self-doubt to assertive authenticity underscores the power of resilience, deep relationships, and unwavering purpose, revealing the transformative impact of staying true to oneself amidst adversity****. In the structured corridors of corporate America,* ***I once believed my identity was to be adjusted and reshaped according to prevailing standards and norms****. Those standards and norms weren't set by just one individual, but rather* ***the collective voice of a system steeped in its own definitions of success—a voice predominantly belonging to White men*** *[...] Over time, the muted voice from the beginning of my journey became an* ***assertive, resonant declaration of my identity. No longer muffled by societal expectations****, it proclaimed the* ***power of authenticity****. In the vast expanse of voices, I found my unique tone, not by blending in, but by* ***standing out, firmly rooted in the knowledge of who I truly am*** *[...] Embracing my imperfections has also been liberating. Granting myself the same grace I offered to others enabled me to rise above oppressive situations.* ***Metacognition became a tool, helping me reclaim my space*** *and moderate my reactions, especially during instances of micro and macroaggressions. The power I felt surging within was palpable, propelling me* ***to redefine well-being on my terms*** *[...] My experiences taught me that authentic leadership doesn't involve molding oneself into pre-existing templates, but instead, requires* ***an unwavering dedication to one's true self****.—Simmons*

I would define my leadership style as ***human-centered****. I also consider myself to be a* ***quiet leader****. For human centered leadership, I apply three principles: Empathy, Vulnerability, Humility. Having grown up how I did and my experiences in undergraduate school, which was full of male chauvinism and socio-economic divides, I was* ***keenly aware of the inequities that exist for women and impoverished communities*** *[...]* ***I practice great humility in my leadership****. As an introvert, I have come to appreciate the importance of* ***silence (of your soul) in leadership*** *and how* ***introverts can be more effective leaders than the normative expectation of loud-mouthed outspoken aggressive leaders' engineers typically tend to value****. Finally,* ***I don't assume that I am the holder of power or knowledge but rather that we are all in this together****.—Villanueva Alarcón*

Stemming from their intersecting identities, all were discounted in some way despite performing required job tasks with little to no institutional support. Both implicitly and explicitly, the authors were told that something about their identity or personhood conflicted with the engineering cultural norms. This is an EI example because it operationalizes the notion that different inherently implies deficit. Based on our intersecting identities and lived experiences, our leadership approach has to look different than our White male, heterosexual, middle-class counterparts. Our invisible leadership (Brisbane et al., 2024) is often overlooked, or the overall impact is minimized or diminished as a standard. Yet, each author provided examples of going beyond the required task or added burdens to actually "get her done."

Theme 4: Creating Legacies in Academic Leadership in Engineering
All the authors recognized that their intersecting identities and backgrounds and their countering of the epistemic injustices experienced created opportunities to sow and sustain legacies in academic leadership in engineering. Many indicated having experienced biases, discrimination, harassment, bullying, and retaliations to being authentic to their identities and views of leadership in the future. They were aware of being at the crossroads of change where they were positioned to both fight the fight and live the life they would like to live (whether personal or professional). All of them understood that the battle would not be without scarring but that the outcome would be one of much beauty, strength, and genuineness.

I have worked as a founding faculty at two predominantly white institutions and ***witnessed firsthand the conscious and unconscious biases that exist—I faced exclusion, bullying,***

> ***harassment, discrimination, and retaliation*** *at both of these universities in being a* ***non-US born, woman, engineering leader****. My* ***identity as an Engineer was actually strengthened as a result*** *of being the Founding Chair of Wake Forest Engineering. This is an outcome I never would have expected but* ***the responsibility and advocacy role I had in building from scratch the new engineering program strengthened my identity of being an Engineer*** *(in part because being an Engineer was a minority identity within Wake Forest University)[…] I walked into an institution that not only lacked ethnic and racial diversity, but lacked* ***structures, policies, and procedures that were equitable and just . In contrast, the National Science Foundation as an organization has been a higher functioning and more ethical organization than any academic institution, I have been a part of, and it is because of the diversity of its workforce and the strong structures and policies of accountability against bias, harassment, and discrimination. I believe everyone needs to see themselves as a leader and an agent of change within any organization they are a part of.*** *An administrative title does not make a leader,* ***a person's actions make a leader and all of us can be leaders when we conduct ourselves with the highest standards of character****. We truly need more leaders who bring not only competence but character to any role.—Pierrakos*

> *In my* ***zealous attempts to help and serve, I've sometimes overextended, occasionally facing rejection or worse****. It was a profound realization that* ***not all aid is perceived as helpful****, and sometimes,* ***in attempting to assist, I was serving my own interests over others'****. It became my imperative to continually reflect and ask, "What does help look like for you?" ensuring that support is aligned with the genuine needs of those I sought to serve […] Leadership, at its core, is not about titles or positions, but* ***the influence and impact one leaves behind****. The profound lessons I've* ***gleaned from the remarkable figures in my life have not only shaped my personal worldview but have also informed my approach to leadership*** *[…] I aim to* ***lead by example, showcasing that it's possible to be firm yet kind, to be decisive yet open to feedback, and to lead with both heart and mind****. It's my hope that as I continue to evolve as a leader, I* ***can inspire others to embrace their unique strengths, lean into their values, and leave an indelible mark on the world around them****. After all,* ***leadership is not about a moment in time, but a legacy we leave for generations to come****.—Simmons*

Each author expressed their intention to leave the field better than what they inherited by living the concept of Sankofa (Temple, 2010) where you pull others along as you climb. This phenomenon illustrates EI in two ways: (1) it is not a challenge that our counterparts have to deal with or manage, the requirement to pave the way for others, and (2) also this requirement supports multiple characteristics of the White supremacist culture (Okun & Jones, 2000): "defensiveness" to even consider race or raced history and "only one right way" where if it does not look consistent with dominant cultural way, then it is wrong. Therefore, each participant not only had to be a living example to model professional behaviors but also typically had to mentor to teach their existing leadership how to support them.

11.6 Strategies

11.6.1 Becoming Diverse and Authentic Academic Leaders in Engineering

All authors willingly shared some of the strategies that they have learned and used as they have transitioned into different leadership roles: from thought leaders to associate chair to founding chair to associate dean to director. With different stages of their career progression, their views and strategies about academic leadership have evolved and yielded new wisdom for the next generation to come. Rather than parse out the list of suggestions, the authors decided to maintain the authentic voices as they tackled their epistemic injustices.

- *A recommendation I would share is to document strategies as they navigate it. Documenting the strategies may involve reaching out to internal and external mentors you trust and asking them how they have navigated similar scenarios [...].*
- *[...] You cannot compromise who you are and your leadership style for anyone, especially in the engineering field. So, bring your true self to the space, and if they do not value it, it may not be worth your time and effort.*
- *Embrace every lesson, cherish every relationship, and always seek feedback to continue growing. In understanding our past and valuing our present, we can forge a future filled with purpose, gratitude, and unwavering faith.*
- *Ultimately, my leadership journey has been about transcending barriers and creating pathways for others to do the same. It is not about wielding power but about empowering. It is not about individual success but collective upliftment. My commitment to lead is rooted in my experiences, and it is this commitment that drives me to make a difference every single day. As I lead, I carry with me the lessons of the past, ensuring that no voice, no matter how soft, goes unheard.*
- *In the multifaceted realm of engineering, leadership is not just about holding authoritative roles but understanding oneself, embracing authenticity, and addressing the nuanced challenges posed by intersecting identities.*
- *For those with intersecting identities aspiring to become academic leaders in engineering, it is paramount to navigate through these complex terrains by drawing from both lived experiences and strategic insight.*
- *Leadership is less about the roles we occupy and more about personal growth and self-awareness. I have come to embrace my imperfections, a vital aspect of leadership, offering myself the same grace I extend to others. This journey towards self-understanding has not only made me a better mentor and leader but also fortified my resilience.*
- *Specifically, I purposefully select my teams and colleagues that feed my soul/spirit, share my values with the community, and have my back. Secondly, I maintain my emotional integrity and balance by naming the subtle racism and the overt racism that I experience and witness. Next, I actively implore all to take responsibility for inclusiveness in STEM and mitigate historical and systemic racism that supports the White supremacy culture. Finally, to keep my soul intact and maintain balance with beliefs and work [...].*
- *Develop strategies for daily, weekly, semester, and yearlong plans to keep your soul intact.*
- *Attack behaviors or policy, not person; practice how you want to respond to discriminatory events by planning your response or fire drill (make a plan for an emergency).*
- *I have learned that we have multiple identities and that the environments we are in make some of these identities more salient than others. These identities have status too [...]. Self-awareness and adaptability are fundamental, but staying true to who we are is a must.*

- *I advise colleagues to pay attention to the broad diversity (gender, race, ethnicity) of the organizations they join and to pay attention to the policies and procedures in place against bias, harassment, and discrimination. If you do not see yourself represented in the leadership of the organization you are joining, think twice about joining because making positive change happen will be a struggle.*

11.7 Conclusions

In conclusion, we described how each theme illuminates the aspects of epistemic injustice within the context of engineering and is encapsulated in each author's journey to becoming an effective intersectional academic leader. Overall, this chapter exposed the intersectional stories and dimensions of personal upbringings that include political, cultural, economic, and other legalities that situate the intersectional identities of the authors. Using an autoethnographic approach, the authors shared their stories, identified points of intersection, and used them to identify the commonalities and differences in experiences. From that, strategies around academic leadership in engineering were synthesized and shared for the reader to explore. While not prescriptive, the authors hope to communicate the value added by women who are academic leaders with diverse intersectional identities in engineering spaces.

Acknowledgments This material is based upon the work supported by the National Science Foundation under Award No 2417100 (NSF IPA Agreement for Olga Pierrakos). Any opinions, findings, and conclusions or recommendations expressed in this material are those of the author(s) and do not necessarily reflect the views of the National Science Foundation.

We would like to acknowledge Dr. Cross's graduate student, Kaitlyn Thomas for her thoughtful discussion on epistemic injustice in this chapter.

References

Beddoes, K., & Borrego, M. (2011). Feminist theory in three engineering education journals: 1995–2008. *Journal of Engineering Education, 100*(2), 281–303. https://doi.org/10.1002/j.2168-9830.2011.tb00014.x

Bhakuni, H., & Abimbola, S. (2021). Epistemic injustice in academic global health. *The Lancet Global Health, 9*(10), e1465–e1470. https://doi.org/10.1016/S2214-109X(21)00301-6

Bowleg, L. (2008). When Black+ lesbian+ woman≠ Black lesbian woman: The methodological challenges of qualitative and quantitative intersectionality research. *Sex Roles, 59*, 312–325.

Brisbane, J. M., London, J., Perry, L., Coley, B., Thomas, K., Greene, M., & Maitra, D. (2024). The Added Value of Black Engineering Students' Participation in Identity-based Organizations: A Systematic Literature Review. *Studies in Engineering Education, 5*(1), 1–23. https://doi.org/10.21061/see.79

Callimachi, R. (2021, Nov. 17). Lost lives, lost culture: The forgotten history of Indigenous boarding schools. Retrieved August 2, 2024, from New York Times. https://www.nytimes.com/2021/07/19/us/us-canada-indigenousboarding-residential-schools.html

Campbell, J. D., Assanand, S., & Di Paula, A. (2003). The structure of the self-concept and its relation to psychological adjustment. *Journal of Personality, 71*, 115–140.

Chang, H. (2016). Autoethnography in health research: Growing pains? *Qualitative Health Research, 26*, 443–451. https://doi.org/10.1177/1049732315627432

Chang, H., Longman, K. A., & Franco, M. A. (2014). Leadership development through mentoring in higher education: A collaborative autoethnography of leaders of color. *Mentoring & Tutoring: Partnership in Learning, 22*(4), 373–389. https://doi.org/10.1080/13611267.2014.94573

Collins, P. H. (2015). Intersectionality's definitional dilemmas. *Annual Review of Sociology, 41*(1), 1–20.

Collins, P. H., & Bilge, S. (2016). *Intersectionality—Key concepts*. Polity Press.

Costello, J., Feller, J., & Sammon, D. (2016). On the road to trusted data: An autoethnography of community governance and decision-making. *Journal of Decision Systems, 25*(sup1), 182–197. https://doi.org/10.1080/12460125.2016.1187413

Crenshaw, K. (1990). Mapping the margins: Intersectionality, identity politics, and violence against women of color. *Stanford Law Review, 43*, 1241–1299. https://doi.org/10.2307/1229039

Crenshaw, K. W. (1989). *Demarginalizing the intersection of race and sex: A black feminist critique of antidiscrimination doctrine, feminist theory, and antiracist politics* (pp. 139–167). University of Chicago Legal Forum.

Crenshaw, K. W. (2017). *On intersectionality: Essential writings*. The New Press.

Cross, K. J., Gaskins, W. B., & Coley, B. C. (2022). We cannot address what we do not acknowledge. *Understanding and Dismantling Privilege, 12*(1), 68–85.

Dember, W. N. (1974). Motivation and the Cognitive Revolution. *The American Psychologist, 29*, 161–168.

Diversity and Women in STEM: Women, Minorities, and Persons with Disabilities. (30 January, 2023). U.S. National Science Foundation). Retrieved August 2, 2024, from https://new.nsf.gov/news/diversity-and-stem-2023

Dotson, K. (2014). Conceptualizing Epistemic Oppression. Social Epistemology, 28(2), 115–138. https://doi.org/10.1080/02691728.2013.782585

Du Bois, W. E. B. (1897). Strivings of the negro people. *Atlantic, 80*, 194.

Du Bois, W. E. B. (1903). *The souls of black folk*. (rpt. (Vol. 1989, p. 5). Penguin.

Ellis, C., Adams, T. E., & Bochner, A. P. (2011). *Autoethnography: An overview* (pp. 273–290). Historical Social Research/Historische Sozialforschung. Retrieved from https://www.jstor.org/stable/23032294

Fricker, M. (2000). Feminism in epistemology: Pluralism without postmodernism. In The Cambridge companion to feminism in philosophy (pp. 146–165). Cambridge University Press.

Fricker, M. (2007). Epistemic Injustice: Power & the Ethics of Knowing. Oxford University Press.

Judson, E., Ross, L., Hjelmstad, K. D., Krause, S. J., Ankeny, C. J., Culbertson, R. J., & Middleton, J. A. (2017). Examination of implicit gender biases among engineering faculty when assigning leadership, research, and service roles. *ASEE Annual Conference and Exposition, Conference Proceedings, 2017-June*. https://doi.org/10.18260/1-2-28314

Lackey, J. (2006). Learning from Words. *Philosophy and Phenomenological Research, 73*(1), 77–101.

Lackey, J. (2008). A Justification is View of Disagreement's Epistemic Significance. *Proceedings of the XXII World Congress of Philosophy, 53*, 145–154. https://doi.org/https://doi.org/10.5840/wcp2220085323.

Manis, M. (1977). Cognitive social psychology. *Personality and Social Psychology Bulletin, 3*, 550–566.

Martin, J., Desing, R., & Borrego, M. (2022). Positionality statements are just the tip of the iceberg: Moving towards a reflexive process. *Journal of Women and Minorities in Science and Engineering, 28*(4).

McGee, E., Cox, M. F., Main, J. B., et al. (2024). Wage disparities in academia for engineering women of color and the limitations of advocacy and agency. *Research in Higher Education, 65*(5), 914–942. https://doi.org/10.1007/s11162-023-09766-3

McKinnon, R. (2016). Epistemic Injustice. *Philosophy Compass, 11*(8), 437–446. https://doi.org/10.1111/phc3.12336

McLoughlin, L. A. (2005). Spotlighting: Emergent Gender Bias in Undergraduate Engineering Education. *Journal of Engineering Education, 94*(4), 373–381.

Mejia, J., Alarcón, I. V., Mejia, J., & Revelo, R. (2022). Legitimized tongues: Breaking the traditions of silence in mainstream engineering education and research. *Journal of Women and Minorities in Science and Engineering, 28*(2).

Okun, T., & Jones, K. (2000). White supremacy culture. In *Dismantling racism: A workbook for social change groups*.

Ong, M., Jaumot-Pascual, N., & Ko, L. T. (2020). Research literature on women of color in undergraduate engineering education: A systematic thematic synthesis. *Journal of Engineering Education, 109*(3), 581–615. https://doi.org/10.1002/jee.20345

Pogrebna, G., Angelopoulos, S., Motsi-Omoijiade, I., Kharlamov, A., & Tkachenko, N. (2024). The impact of intersectional racial and gender biases on minority female leadership over two centuries. *Scientific Reports, 14*(1), 111.

Revelo, R. A., Mejia, J. A., & Villanueva, I. (2017, June). Who are we? Beyond monolithic perspectives of Latinxs in engineering. In *2017 ASEE Annual Conference & Exposition*.

Rubin, D. C. (2020). Self-concept focus: A tendency to perceive autobiographical events as central to identity. *Journal of Applied Research in Memory and Cognition, 9*(4), 576–586.

Rubin, D. C., Berntsen, D., Deffler, S. A., & Brodar, K. (2019). Self-narrative focus in autobiographical events: The effect of time, emotion, and individual differences. *Memory and Cognition, 47*, 63–75.

Saldaña, J. (2016). *The Coding Manual for Qualitative Researchers*. SAGE Publications, Incorporated.

Secules, S., McCall, C., Mejia, J. A., Beebe, C., Masters, A. S., Sánchez-Peña, L. M., & Svyantek, M. (2021). Positionality practices and dimensions of impact on equity research: A collaborative inquiry and call to the community. *Journal of Engineering Education, 110*(1), 19–43.

Settles, I. H., Jones, M. K., Buchanan, N. C. T., & Dotson, K. (2021). Epistemic Exclusion: Scholar(ly) Devaluation That Marginalizes Faculty of Color. *Journal of Diversity in Higher Education, 14*(4), 493–507. https://doi.org/10.1037/dhe0000174

Singh, I. (27 November, 2020). By the Numbers: Women in STEM: What do the statistics reveal about ongoing gender disparities? Yale Scientific. Retrieved from https://www.yalescientific.org/2020/11/by-the-numbers-women-in-stem-what-do-the-statistics-reveal-about-ongoing-gender-disparities/

Sochacka, N. W., Guyotte, K. W., & Walther, J. (2016). Learning together: A collaborative autoethnographic exploration of STEAM (STEM+ the Arts) education. *Journal of Engineering Education, 105*(1), 15–42. https://doi.org/10.1002/jee.20112

Solórzano, D. G., & Yosso, T. J. (2001). Critical race and LatCrit theory and method: Counter-storytelling. *International Journal of Qualitative Studies in Education, 14*(4), 471–495.

SWE (2023, January 1). *Tenure/Tenure-Track Faculty Levels*. Society of Women Engineers. Retrieved August 2, 2024, from https://swe.org/research/2024/tenure-tenure-track-faculty-levels/

Temple, C. N. (2010). The emergence of Sankofa practice in the United States: A modern history. *Journal of Black Studies, 41*(1), 127–150.

Williams, E. N., & Morrow, S. L. (2009). Achieving trustworthiness in qualitative research: A pan-paradigmatic perspective. *Psychotherapy Research, 19*(4–5), 576–582. https://doi.org/10.1080/10503300802702113

Women, Minorities, and Persons with Disabilities in Science and Engineering. (29 April, 2021). U.S. National Science Foundation). Retrieved August 2, 2024, from https://ncses.nsf.gov/pubs/nsf21321/report/persons-with-disability

Woolford, A. (2015). *This benevolent experiment: Indigenous boarding schools, genocide, and redress in Canada and the United States*. U of Nebraska Press.

Chapter 12
Women in Engineering Beyond the Gender Binary

Kelly J. Cross and Donna Riley

12.1 Introduction

This book is about women in engineering. Using the label woman carries with it a set of meanings, assumptions, and definitions that we each may apply differently. Who is included, and who is excluded? Who is empowered by the use of this term? Whose experience is centered, and who may (intentionally or unintentionally) be left in the margins, or excluded entirely? For example, do we mean only cis-hetero women? Do we only imagine white women or white women as the norm in engineering? Do we include experiences of older women, younger women, immigrant women, economically disadvantaged women, parents and nonparents, transwomen, and disabled women, or are those women less visible to us? Are non-binary people who do not use the label woman for themselves included? Are transmen? Is the category established based solely on identity or on other factors related to upbringing, biology, hormones, or other experiences?

We seek here to both expand our collective thinking of what we might mean by "women in engineering" and expand our understanding of gender from an intersectional perspective, employing lenses that recognize both that women embody intersecting identities *and* that those identities create specific experiences of power in engineering, throughout academia, and beyond.

K. J. Cross
Georgia Tech and Emory University, Atlanta, GA, USA
e-mail: kelly.cross@bme.gatech.edu

D. Riley (✉)
University of New Mexico, Albuquerque, NM, USA
e-mail: riley1@umn.edu

J. P. Carpenter (ed.), *Women in Academia*, Women in Engineering and Science,
https://doi.org/10.1007/978-3-031-97938-5_12

12.2 Expanding Definitions

Gender identities are fluid and ever-evolving, both individually and collectively (Butler, 1990). Donna, for example, has gone in her lifetime from mostly conforming to her mother's ideal of femininity, to using creative language that embodied feminist critique (e.g., Cameron, 1990), and to adopting a genderqueer label in the early 2000s (and suggesting, loosely after Kate Bornstein (1994) and others, that women in engineering may by definition be queering or trans-gressing gender if engineering is indeed a traditionally male profession) (Riley, 2008). Today, using a combination of she/they pronouns, she recognizes her femme-ish genderqueer sensibility while seeking to hold space for all gender identities in the engineering community. Kelly was socialized about gender through the Black church. Her life, intellectual, and spiritual journey has expanded her understanding and critique of socially accepted norms about gender. One critical event occurred when a work colleague introduced Kelly to the concept of a dual spirit based on the Lakota Indigenous tradition. Kelly now identifies as a dual spirit. Furthermore, in a lesbian community, Kelly learned about she+ as designation of an evolving definition and understanding of the female gender or womanism. Kelly also uses she/they pronouns but will occasionally refuse to specify a pronoun to hold space for those who are not free to specify their pronouns or their dynamic gender expression. Allowing the spectrum of women to define their gender for themselves is a step towards inclusive excellence and removing the harmful judgements we all learned in our hegemonic patriarchal societal training, aka upbringing and education.

12.3 Inclusive Spaces

Inclusive spaces, then, must consider gender identities beyond dichotomies of male and female—and beyond male, female, and non-binary. It is especially important not to **police others' gender identities**, not to seek to put someone in a box or judge their presentation or expression of their identity. We must recognize identity as not static, but malleable, fluid, and changing with context. As lifelong learners, we live into the meaning of our identities and embody new realities of gender even if our labels and pronouns do not keep up with our new realities.

12.4 Intersectionality

In Crenshaw, 1989, legal scholar Kimberlé Crenshaw coined the term intersectionality, to account for the specific lived experiences of women of color. She noted that Black women's experience under the law could never be understood merely as a combination of Black men's experiences and women's experiences in general, but rather Black women's experiences needed to be understood on their own terms, as they were subject to particular power dynamics and outcomes all their own. (Bowleg, 2008; Bowleg & Bauer, 2016) described the intersectional nature of gender and how other identity dimensions can affect and be affected by gender and gender expression.

In this way, the category *woman* cannot be reduced to a universal common experience but must be understood as containing interrelated identities and experiences. One cannot consider sexism in a vacuum but instead must seek to learn how racism works simultaneously to shape the experiences of women of color. This work has been extended to incorporate other forms of oppression such as ableism or heterosexism, thus developing a deepening understanding of the experiences of queer women of color, or disabled women of color, who each experience power and oppression in ways that are again distinct from non-queer or able-bodied women of color. Recent engineering education research problematized the binary of gender and the simplification within research (Haverkamp et al., 2020). Layering in engineer as an additional intersecting identity, and studying intersectional power (Collins & Bilge, 2020) within engineering education, is important work only beginning to be considered recently (Perkins et al., 2019; Svihla et al., 2023; True-Funk et al., 2021).

12.5 Conclusion

We close by noting that one could further problematize the concept of *engineering* and its power relations, but that discussion is left to the reader.

References

Bornstein, K. (1994). *Gender Outlaw: On Men, Women, and the Rest of Us*. Routledge.

Bowleg, L. (2008). When Black+ lesbian+ woman≠ Black lesbian woman: The methodological challenges of qualitative and quantitative intersectionality research. *Sex Roles, 59*, 312–325.

Bowleg, L., & Bauer, G. (2016). Invited reflection: Quantifying intersectionality. *Psychology of Women Quarterly, 40*(3), 337–341.

Butler, J. (1990). *Gender trouble: Feminism and the subversio of identity*. Routledge.

Cameron, D. (Ed.). (1990). *Feminist critique of language: A reader*. Routledge.

Collins, P. H., & Bilge, S. (2020). *Intersectionality*. John Wiley & Sons.

Crenshaw, K. (1989). Demarginalizing the intersection of race and sex: A black feminist critique of antidiscrimination doctrine, feminist theory and antiracist policies. *University of Chicago Legal Forum, 1*, 139–167.

Haverkamp, A., Bothwell, M., Montfort, D., & Driskill, Q.-L. (2020). Calling for a paradigm shift in the study of gender in engineering education. *Studies in Engineering Education, 1*(2), 55–70. https://doi.org/10.21061/see.34

Perkins, H. L., Bahnson, M., Tsugawa-Nieves, M. A., Satterfield, D. J., Parker, M., Cass, C., & Kirn, A. (2019). An intersectional approach to exploring engineering graduate students' identities and academic relationships. *International Journal of Gender, Science and Technology, 11*(3), 440–465.

Riley, D. M. (2008). LGBT-friendly workplaces in engineering. *Leadership and Management in Engineering, 8*, 19–23. https://doi.org/10.1061/(ASCE)1532-6748(2008)8:1(19)

Svihla, V., Davis, S. C., & Kellam, N. N. (2023). The TRIPLE change framework: merging theories of intersectional power, learning, and change to enable just, equitable, diverse, and inclusive engineering education. *Studies in Engineering Education, 4*(2), 38–63. https://doi.org/10.21061/see.87

True-Funk, A., Poleacovschi, C., Jones-Johnson, G., Feinstein, S., Smith, K., & Luster-Teasley, S. (2021). Intersectional engineers: Diversity of gender and race microaggressions and their effects in engineering education. *Journal of Management in Engineering, 37*, 04021002. https://doi.org/10.1061/(ASCE)ME.1943-5479.0000889

Chapter 13
Women in Engineering: The Contribution of Women's Colleges

Donna Riley

13.1 Introduction

No book on women in engineering would be complete without consideration of engineering programs at women's colleges. In 1999, Smith College was the first to announce its creation of the Picker Engineering Program, which would graduate its first class in 2004 and earn retroactive ABET accreditation in 2005.

President Ruth Simmons provided a clear rationale for its founding. "Engineers design and build much of the human environment. Women must not accept so marginal a role in so important a field" (Clayton, 2000). In 2005, Sweet Briar College would follow suit with its own Engineering Program, graduating its first class in 2009 and receiving ABET accreditation soon after (House, 2011).

It is difficult to systematically gauge the impact of these programs on engineering and on academia. I offer here my personal observations and reflections as a founding faculty member at Smith who served there until 2014.

13.2 Defining the Difference

As the first class of Smith alums recently celebrated their 20th reunion, one can see the impact of these 19 lives in engineering academia, in practice, and in many other fields. There are tenured academics like Caitlyn Butler, Civil Engineering faculty at the University of Massachusetts; Fatima Toor, Electrical Engineering faculty at the University of Iowa; and Cara Stepp, Speech, Language, and Hearing Sciences faculty at Boston University. There are registered professional engineers working in

D. Riley (✉)
University of New Mexico, Albuquerque, NM, USA
e-mail: riley1@umn.edu

J. P. Carpenter (ed.), *Women in Academia*, Women in Engineering and Science,
https://doi.org/10.1007/978-3-031-97938-5_13

industry, and others who have assumed c-suite roles in tech startups. There are those who went on to practice in different professions including business, language and culture, and technology policy. Becky Silverstein went on to graduate from seminary and became the first out trans rabbi to be hired by a Conservative synagogue (Several, 2017).

As part of the founding faculty cohort, we sought to do engineering academia differently. In particular, we shared a desire to enact a different work-life balance in our academic careers, with most of our founding cohort starting families while we were yet establishing the program. The male professor among us who was starting a family alongside female colleagues was married to a female professional engineer who worked full time. Gender inequities in Smith's parental leave policies quickly became apparent, when mothers, but not fathers, were fast-tracked for daycare. Ultimately, this test case, and the advocacy around it, made space for equitable treatment of all parents, regardless of gender.

In these matters as well as in curriculum and pedagogy, we had the support of our founding director, Domenico Grasso, to explore how things could be different at Smith. At the same time, we faced sexism from the rest of the engineering academia, wondering if we were doing "engineering light" because we were a "girls' school." This imposed a certain conformity to standards of "rigor" (and a simultaneous emergent critique) (Riley, 2017).

One of the things commonly observed about women's colleges is that every student leadership position on campus is drawn from the student body: the editor of the school newspaper, the president of the student body, the head of the radio station, and the star athletes—positions frequently awarded to cisgender men on coed campuses. Access to leadership experience is amply available to women, trans, and non-binary students, and the confidence it instills in alums is palpable. It was thus not at all surprising that Smith's Picker Engineering Program sent the first all-women team ever to fly in the NASA reduced gravity flight opportunities program (Weld, 2003).

I found myself surprised by the distribution of engineering disciplinary interest among our students. The founding faculty's focus areas were clustered in subfields that nationally represented the most women: bioengineering (biomechanics, speech, and hearing) and civil/environmental engineering (water quality, hydrology, environmental risk). While there certainly were students interested in those areas, the interests of many of our students more closely resembled engineering students as a whole, with many if not most choosing to concentrate in mechanical and electrical engineering. As with other engineering programs, many went to work in traditional engineering jobs in defense (e.g., United Technologies companies like Sikorsky or Pratt & Whitney and Boeing). Sweet Briar College's list of career placements suggests a similar pattern, with many entering the Department of Defense (Army, Navy, NAVAIR), contractor field (Boeing, Northrop Grumman), and the nuclear power industry (Dominion, Babcock & Wilcox, Southern Nuclear) (Sweet Briar College, 2025). Of course, engineering firms in construction and environmental engineering, or in automotive, consumer products, energy efficiency, and much more, are also represented. But it is a reminder that we sometimes make too much of gender difference in saying that women prefer to work in humanitarian or people-focused areas. It seems to me that the learning environments at Smith and Sweet Briar served to remove at least some of the pressure to focus in more feminized areas of engineering. At the same time, it is notable that some graduates pursue nonprofit work, government consulting, Wall Street, and other fields, as you might expect from a liberal arts education. And once an engineer, always an engineer; these graduates are bringing an engineering education, its perspective, and its value—to new settings.

13.3 Integrative Explorations: Engineering and the Liberal Arts

While Sweet Briar followed Smith's lead, many other women's colleges (notably Wellesley, Mt. Holyoke, Spelman, and Bryn Mawr, all offering transfer "2 + 2" agreements with large engineering programs, but no stand-alone program) did not. I became curious about how it was that Smith was able to establish an engineering program when others chose not to do so. In true liberal arts fashion, I sat in on a women's history course and worked with women and gender studies student Heidi Waugh to explore these questions. In our research in the Sophia Smith Collection and Smith's archives, we learned about Math Professor Alice Dickinson, who built radar systems in World War II and became Smith's Dean of the College in the 1970s (Dickinson, 1961–1977). When Jill Ker Conway assumed the presidency in 1975, Dr. Dickinson led the effort to establish minors in engineering conferred by Smith based on coursework at both Smith and nearby UMass.

This program continued apace for Conway's tenure, but was disrupted by the next president, Mary Maples Dunn, whose vision of the liberal arts did not include engineering (Dunn, 1985). But by the time the program was dismantled in 1991 (Hellman & Merritt, 1991), several faculty members in the sciences had already been hired (and tenured) with both engineering backgrounds and a passion for educating women in engineering. When newly appointed President Simmons invited proposals in the late 1990s for bold new initiatives, these faculty and their colleagues were ready to propose the first stand-alone engineering program at a US women's college. The debates about the role of engineering in a liberal arts college would continue, but at a college established by Sophia Smith "with the design to furnish for my own sex means and facilities for education equal to those which are afforded now in our colleges to young men"; it ultimately fit with Smith's institutional identity (Smith, 1871).

Many of my Smith colleagues and I in turn sought to bring the best of Smith (and liberal arts colleges more generally) into engineering. The collisions of engineering with the humanities and social sciences shaped engineering for the better, producing for example a considerable number of engineers who studied abroad for a full year, a large cohort who enjoyed the opportunity to live and think together with students across disciplines, and many who learned to view engineering differently through Smith's curriculum and pedagogies. I believe that Smith and Sweet Briar—and the other engineering programs at liberal arts colleges like Swarthmore, Lafayette, Union College, and others—continue to make these kinds of differences in engineers' formation.

My own entrance to the field of engineering education occurred in a way that might only have happened at Smith. I was struggling with hierarchical classroom pedagogies in my thermodynamics course, which produced uneven engagement from students by race and class, among other characteristics. My colleague in sociology at another liberal arts college (Grinnell) suggested that I read bell hooks's (1994) *Teaching to Transgress*. It changed everything about how I approached my classroom and the material itself. I began decentering Western civilization in the engineering curriculum, employing feminist, critical, and culturally responsive pedagogies, and teaching the material with its sociotechnical context. I did so with the help of colleagues across campus and internal endowments that supported my work with students to develop my courses in these ways. When I embarked on integrating Foucault's (1980) discussion of Truth and Power in science in my thermodynamics course, my students met me there. They knew Foucault, either from classes they had taken or from conversations with housemates. They were willing to consider what it might mean for engineering, because of the broad liberal arts context of their education.

In 2005 or 2006, some Smith students founded a queer engineers group, mostly as a countercultural statement about the odd sort of way that our Society of Women Engineers chapter had grown to be, in their view, a kind of departmental pep club. They were in a lighthearted way challenging norms around gender, questioning what it might mean to be doing engineering at a women's college, and mostly making room for a variety of identity positions in relation to the idea of women in engineering. That this happened organically in the first few years of our program continues to delight me. These students were uncannily savvy about fluidity around sexual and gender identities, and they understood the power and politics of the bold proposition Smith's Engineering Program made to the world. I continue to be grateful for the lessons I learned from them.

13.4 Conclusion

My career has taken me away from Smith College to the National Science Foundation, and then two large land-grant universities, and now to a medium-sized public flagship where I serve as Dean of Engineering. I know that Smith and Sweet Briar no longer garner the attention they once did in national conversations about engineering and engineering education, even in conversations on women in engineering. But the dozens of female, trans, and non-binary engineering graduates these programs turn out each year continue to shape the engineering profession, engineering academia, and many other domains, in the ways I have attempted to describe here, and no doubt in many ways unknown to me. Their leadership is an inspiration to me, and I strive to amplify their stories as often as I can. It is my hope that you will have the opportunity to encounter a Smithie or Sweet Briar engineer in your work and feel the seismic shift that these programs, and their alums, have created in engineering and beyond.

References

Clayton, M. (January 4, 2000). Engineering 100: No Men Allowed. The Christian Science Monitor. Retrieved from https://www.csmonitor.com/2000/0104/p14s1.html

Dickinson, A. (1961–1977). Alice Dickinson Papers. Sophia Smith Collection, Smith College. Retrieved from https://findingaids.smith.edu/repositories/4/resources/1521

Dunn, M. M. (1985). The inaugural address. *The Smith Alumnae Quarterly, 77*((1)), 8–9.

Foucault, M. (1980). "Truth and power," Alessandro Fontana and Pasquale Pasquino, interviewers. In C. Gordon (Ed.), *Power/knowledge: Selected interviews and other writings 1972–1977* (pp. 131–133). New York, NY.

Hellman, K., & Merritt, R. (1991). Memo to Mary Maples Dunn on behalf of the Engineering Advisory Committee. In Possession of the Author.

hooks, B. (1994). *Teaching to transgress; Education as the practice of freedom*. Routledge.

House, M. (November, 2011). *Engineering program moves into accreditation: Sweet briar college celebrates recently acquired accreditation status*. Lynchburg Business. Retrieved from https://web.archive.org/web/20150402131415/; http://www.lynchburgbusinessmag.com/engineering-program-moves-into-accreditatio-sweet-briar-college-celebrates-recently-acquired-accreditation-status/

Riley, D. (2017). Rigor/Us: Building boundaries and disciplining diversity with standards of merit. *Engineering Studies, 9*(3), 249–265. https://doi.org/10.1080/19378629.2017.1408631

Several, M. (February, 2017). Our history (Pasadena Jewish Temple & Center). Retrieved from https://www.pjtc.net/our-history

Smith, S. (1871). Last will and testament of Miss Sophia Smith, Late of Hatfield, Mass., Deceased; Trumbull & Gere.

Sweet Briar College, After Sweet Briar. (2025). [Engineering Science Webpage]. Retrieved from https://www.sbc.edu/academics/engineering/after-sweet-briar/

Weld, E. S. (2003). *Engineering students risk nausea for NASA*. NewsSmith. Retrieved from https://www.smith.edu/newssmith/spring2003/nasa.php

Chapter 14
Advocates and Allies Programs

Roger Green and Canan Bilen-Green

14.1 Introduction

In the early 2000s, a group of mostly women faculty met to discuss, plan, and implement strategies to improve equity at North Dakota State University (NDSU). Throughout these discussions and activities, one thing became glaringly clear: few men participated or engaged in gender equity efforts. This issue was further highlighted in November 2007 when a *Chronicle of Higher Education* article ranked NDSU near the bottom among 222 doctoral institutions for the percentage of women faculty. This public acknowledgement sparked broader campus engagement and increased awareness of the need for equity efforts, leading to campus-wide events and deeper conversations on gender equity. During one such conversation, a senior colleague noted that "men will need to be involved." Unless addressed, the under-engagement of men represented a significant barrier to truly transform the institution to be more gender equitable. These observations led to a relatively simple yet powerful idea: create an intentional program that engages and develops men to better support gender equity. Modeled after successful social justice training programs such as white anti-racism workshops and Safe Zone LGBTQ ally training, the program focuses on actively engaging men in gender equity initiatives. By educating men and encouraging their active participation, the program aims to broaden support for gender equity across the campus, ensuring that both women and men contribute to and benefit from these efforts. A National Science Foundation Institutional Transformation grant provided the resources to officially launch NDSU's Advocates and Allies initiative. Since its inception in 2009, the Advocates and Allies program has resonated with many institutions and organizations, with new Advocates and Allies programs being launched every year.

R. Green (✉) · C. Bilen-Green
North Dakota State University, Fargo, ND, USA

J. P. Carpenter (ed.), *Women in Academia*, Women in Engineering and Science,
https://doi.org/10.1007/978-3-031-97938-5_14

14.2 Rationale

Organizations with healthy commitments to diversity, equity, and inclusion (DEI) outperform their less-committed counterparts. Thus, it is in an organization's best interests to value and grow DEI initiatives and outcomes. Climate is of particular importance to organizational success and is largely a product of the people and policies that are in place. In academia, gender inequity is a persistent problem, especially in the science, technology, engineering, and mathematics (STEM) disciplines, and poor climate is a primary barrier to improvement. As all members of a unit or organization contribute to climate, it follows that all, or at least most, members need to be involved in DEI initiatives and solutions. Too often, however, responsibility for gender equity efforts disproportionately falls onto women. Without the involvement of majority groups, improving organizational climate becomes difficult if not impossible. Advocates and Allies (A&A) programs seek to intentionally engage men as active supporters of gender equity. By involving men, Advocates and Allies programs increase an organization's capacity for meaningful and lasting change. Furthermore, and somewhat paradoxically, men can sometimes advocate for gender equity in ways that women cannot, perhaps, in part, since men are less likely to be perceived as operating from a position of self-interest (Bilen-Green et al., 2013).

The need to engage men in gender equity work does not, in itself, explain why formal programs, such as Advocates and Allies, may be necessary. First, when an organization is dominated by one group, that group is often unaware of the ways in which climate serves that group and disadvantages others. Since men generally do not suffer from the primary disadvantages of gender inequity, they can function adequately despite existing inequities. Research also identifies key forces that undermine men's engagement as gender equity advocates, including apathy, fear, and lack of knowledge (Prime & Moss-Racusin, 2009). Intentional, structured, formalized programs such as Advocates and Allies help overcome these barriers and provide motivation, knowledge, skills, and pathways for men to better support gender equity within their organization.

14.3 Structure

The mission of Advocates and Allies is threefold: (1) to introduce men to the knowledge, skills, and strategies to effect positive personal, unit, and institutional change; (2) to emphasize men working with other men while maintaining accountability to women; and (3) to build a supportive network of men allies who are committed to gender equity. Figure 14.1 illustrates the typical structure of an Advocates and Allies program (Anicha et al., 2020a). Notice that the Advocates and Allies model allows for two levels of participation: Allies participate in Ally workshops and generally focus on individual actions in support of gender equity, while Advocates comprise a more formal group of men on campus who meet regularly, conduct Ally and

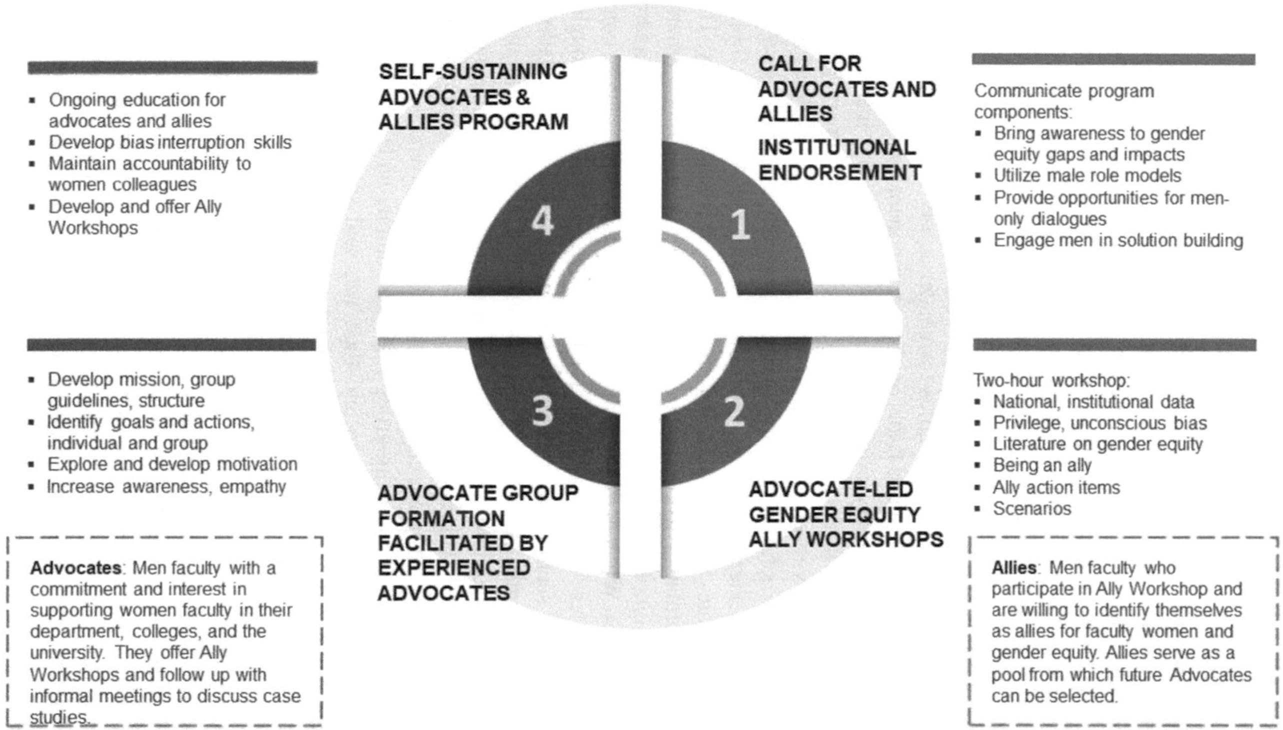

Fig. 14.1 Advocates and Allies program structure (Anicha et al., 2020a)

other trainings, and perform group-level activities in support of gender equity. Allies are key for day-to-day program success, and Advocates are key for program sustainment.

As shown in Fig. 14.1, the first step to establish an Advocates and Allies program involves communication and institutional commitment. A campus group, such as an ADVANCE grant team or DEI office, introduces the campus community to institutional gender equity needs and the Advocates and Allies concept, and it secures institutional and leadership endorsement and support for associated programs and initiatives. Willingness and ability to collect and publicly discuss organizational climate, gender equity gaps, and associated impacts are important. Planning, advertisement, broad recruitment, and registration for an initial series of Ally Workshops are conducted.

In the second step, experienced trainers conduct workshops (usually men-only) to prepare participants as effectives allies for gender equity. Workshops are usually two or more hours in duration and focus on three component areas: (1) introduction and discussion of gender equity issues, experiences, and attitudes, including presentation of institution-specific data; (2) overview of research literature on gender inequity within academia with emphasis on unintended bias; and (3) development of individual skills and actions, so participants can better contribute to a gender-equitable campus. External trainers generally conduct the Ally workshops until institutional Advocates are identified and prepared for this responsibility. The preferred men-only format of workshops helps ensure candid discussion among participants and helps emphasize men's responsibility in understanding and addressing gender equity issues. When workshops allow mixed-gender participation, conversations are generally less candid (as men may feel a need to mask true thoughts and questions) and men seem more inclined to seek explanation and validation from women participants rather than taking responsibility for self-education and reflection. The preferred men-only format also helps promote a functional network of men allies for future education, participation, and action. In fact, initial Ally workshop participants are the prime recruits for application as Advocates.

In the third step, the organizing group makes a campus-wide call and personal invitations to apply as an Advocate and then vets and selects applicants. New Advocates undergo a facilitation process where the group gets to know one another, develops mission and objective statements, considers group structure such as meeting frequency and term commitments, and identifies specific tasks to complete in the upcoming year such as developing and offering Ally workshops. Group mission, objectives, and actions are presented to a women's advocates advisory group (formed by the organizing group) for discussion, feedback, and revision. This interaction between Advocates and representative women on campus is a crucial accountability component. The role of Advocate is more formal than that of Ally, and Advocates are critical to the self-sustainment of an Advocates and Allies program. Usually, Advocates groups are comprised of six to ten men serving 1- or 2-year (often renewable) terms. External facilitators are useful to help first establish Advocates groups. Once an Advocate group is formed and functioning, that group can take responsibility for future recruitment, onboarding, action plan development and refinement, and communication with the women's Advocate advisory group.

The focus of the fourth and final step is on sustainability of the Advocates and Allies program. Ideally, some campus authority, such as a DEI office, suitable administrator, or campus group, monitors that the Advocates group and the women's Advocates advisory group are operational and, if possible, hosts and supports the groups' presence on campus, such as through a suitable and accessible web site. Steps 1–3 in Fig. 14.1 are conducted, usually by the Advocates group, every year, including recording and sharing of annual deliverables and outcomes. Most Advocates programs establish an Advocate coordinator, usually peer-elected from all Advocates, to organize activities, call meetings, and help with other basic logistics.

14.4 Successes, Challenges, and Opportunities

Since the first Advocates program was established in 2009 at North Dakota State University (NDSU), Advocates and Allies workshops and programs have been conducted or established at well over 40 institutions and professional organizations through the direct involvement of NDSU Advocates. Several NSF ADVANCE grants have allowed NDSU to study Ally workshop effectiveness, with high percentages (85% and greater) of participants indicating that they learned new information and gained practical skills to better support gender equity. Program documentation, interviews, and focus groups further confirm positive impacts of Advocates and Allies programs, with participants reporting that they felt more motivated to address gender inequity as a result of their involvement in an A&A program (Anicha et al., 2015; Anicha et al., 2020b). Further, Advocates programs frequently document their yearly accomplishments—including both group and individual accomplishments—providing a direct measure of program impact. Examples include offering workshops, responding to campus needs, supporting existing DEI efforts, reading and discussion groups, individual action plans, policy work, and many others. Overall, there is strong evidence that Advocates and Allies programs provide positive and measurable support for gender equity.

Change is rarely easy, and programs that focus on change often encounter challenges. The challenges experienced by Advocates and Allies programs are often strongly impacted by local factors, but certain challenges occur relatively frequently. It is not uncommon for men who are active in Advocates and Allies programs to experience backlash at various points in their participation. Sustainment of programs is influenced by the presence or absence of program champions. Change is slow, and participant burnout tends to increase with participation duration. Gender is not binary, a fact that can complicate elements of Advocates and Allies programs. Intersectionality of identities can further complicate the role of Advocates and the groups that they serve. While gender equity is the focus of Advocates and Allies groups, the needs and experiences of the BIPOC community and other marginalized groups and identities also need attention and consideration. Differences in cultural and country-of-origin norms can also create challenges. Various configurations, including mixed gender and sessions for those identifying as women, have been attempted to integrate these intersectional realities. However, men-only sessions and discussions continue to be most effective in engaging men, especially STEM men, in gender equity efforts (Anicha et al., 2022). Increasing anti-DEI efforts across the nation can also cause significant complications. Advocates and Allies efforts tend to focus on academic men, but there are other men who make important decisions and strongly influence organizational climate. Advocacy for women leaders can be difficult under the spotlight of attention that leadership brings.

14.5 Conclusion

By design, Advocates and Allies programs are nimble, flexible, and adaptable. Challenges can also be viewed as opportunities, and Advocates and Allies programs can continually adapt and evolve to help meet the needs and overcome the challenges that are present.

References

Anicha, C. L., Bilen-Green, C., & Burnett, A. (2015). Men faculty gender-equity advocates: a qualitative analysis of theory and praxis. *The Journal of Men's Studies, 23*(1), 21–24. Retrieved from https://journals.sagepub.com/doi/10.1177/1060826514561974

Anicha, C. L., Bilen-Green, C., & Burnett, A. (2022). The men need to be involved: a critical frame analysis of gender+ equity narratives. *ADVANCE Journal, 3*(1) Retrieved from https://www.advancejournal.org/article/33632-the-men-need-to-be-involved-a-critical-frame-analysis-of-gender-equity-narratives

Anicha, C. L., Bilen-Green, C., Burnett, A., Green, R., & McGeorge, C. R. (2020a). Just UnDo it: Men faculty addressing gender inequities in academia. *ADVANCE Journal, 1*(2). https://doi.org/10.5399/osu/ADVJRNL.1.2.4

Anicha, C. L., Bilen-Green, C., & Green, R. (2020b). A policy paradox: Why gender equity is men's work. *Journal of Gender Studies, 29*(7), 847–851. https://doi.org/10.1080/09589236.2020.1768363

Bilen-Green, C., Carpenter, J., Doore, S., Green, R., Horton, K., Jellison, K., Latimer, M., Levine, M., & O'Neal, P. (2015). Implementation of advocates and allies programs to support and promote gender equity in academia. In *2015 ASEE Annual Conference and Exposition, Seattle, WA*. Retrieved from https://peer.asee.org/implementation-of-advocates-and-allies-programs-to-support-and-promote-gender-equity-in-academia

Bilen-Green, C., Green, R., McGeorge, C., Anicha, C., & Burnett, A. (2013). Engaging male faculty in institutional transformation. In *Proceedings of the American Society for Engineering Education Annual Conference & Exposition, Atlanta, GA*. Retrieved from https://sftp.asee.org/engaging-male-faculty-in-institutional-transformation

Prime, J., & Moss-Racusin, C. A. (2009). Engaging men in gender initiatives: what change agents need to know. *Research Reviews*. Retrieved from https://www.catalyst.org/wp-content/uploads/2019/01/Engaging_Men_In_Gender_Initiatives_What_Change_Agents_Need_To_Know.pdf

Chapter 15
"But You're Not an Engineer": Exploring the Role of Women Collaborators in Engineering Education Research

Rebecca Brent

15.1 Introduction

My studying education in college, teaching kindergarten and grades 1–4, earning a doctorate in education, and early academic career occupied 19 years of my adult life. During those years, I never had a single thing to do with engineering and did not envision ever having one. And yet, in my subsequent 30 years, engineering education has been my primary professional occupation. In collaboration with engineering educators, I have given a total of 440 teaching workshops on campuses and at conferences around the world and still occasionally give them; codirected faculty development activities for 5 years in a consortium of eight colleges of engineering; and codirected the National Effective Teaching Institute, an annual 3-day workshop for engineering educators given for 25 years under the auspices of the American Society for Engineering Education. I was program evaluator for an NSF ADVANCE grant with female STEM faculty and faculty of color and program evaluator for a multi-site NSF AGEP grant with STEM faculty and graduate students of color. I also coauthored with an engineering professor *Teaching and Learning STEM: A Practical Guide* (Felder & Brent, 2024).

All of these multidisciplinary team-based STEM research projects, faculty development activities, and writing experiences have given me a unique perspective on the phenomenon of non-STEM educators working in STEM education. From early on, I found working in engineering education rewarding in numerous ways. For one thing, I learned a great deal about important things that my education and experience had never taught me, and I believe that my education and experience allowed me to make useful contributions that most engineers were unlikely to think of. Also, I thoroughly enjoyed working with audiences filled with smart, capable people who were eager to learn how to improve what they were doing in the classroom.

R. Brent (✉)
Education Designs, LLC, Chapel Hill, NC, USA

J. P. Carpenter (ed.), *Women in Academia*, Women in Engineering and Science,
https://doi.org/10.1007/978-3-031-97938-5_15

There were also distinct challenges about being isolated in an unfamiliar and not always accepting environment. For instance, I had to learn a whole new vocabulary, quickly gain respect and credibility in a field in which I had no training or prior experience, and adjust to a broad assortment of personality and communication styles much different from mine. In this chapter, I will briefly relate how and why I made such an abrupt change in my professional environment, provide some details about the benefits and challenges of women in non-STEM disciplines going into traditionally male STEM fields such as engineering, and offer some suggestions regarding how to successfully address the challenges.

15.2 Brief History

My leap into engineering education uncoincidentally occurred a few years after I married Richard Felder, a chemical engineering professor at North Carolina State University who had developed a deep interest in pedagogy since joining the faculty. He had delved into the education literature, tried out a few nontraditional (in engineering) teaching methods such as active and cooperative learning, and found that after some trial and error the techniques worked surprisingly well. He published several papers about his discoveries and started getting invitations to speak about his work at other universities, first in brief seminars and later in one-day and longer workshops. At one point, he had an epiphany. As he puts it, "I thought about the fact that I was running all over telling others how to teach when I had exactly zero training in it myself, and here was my wife with her doctorate in education and many years of teaching experience. Duh!" Rich invited me to co-present at his next workshop, which happened to be in Brazil, which sounded like fun, and I agreed to go. We went, and it went well, and it *was* fun. The seminars, workshops, book, research, and program evaluations followed, sometimes involving Rich and increasingly involving me alone or other collaborators.

15.3 Benefits

Obvious benefits of collaborations on education-related projects between STEM and non-STEM experts are how much better the final product is likely to be and how much additional expertise members of both groups are likely to gain. For example, I am familiar with a longitudinal study of engineering education students that involved an engineering professor as principal investigator and a cognitive psychology professor and a statistics professor as co-investigators. The psychologist helped the engineer select the student outcomes to be assessed (including such things as students' locus of control and self-efficacy in the subjects to be taught) and the appropriate instruments to assess them and also provided assistance with qualitative research techniques unfamiliar to the engineer. The statistician designed and carried out appropriate statistical tests on the quantitative assessment data, and one of her main roles throughout the project was to look at the engineer's drafts of reports and journal articles and gently declare, "You can't say that—the data don't support it!" The three-professor team remained together throughout the 5 years it took to complete the study.

A major benefit to working in a discipline not one's own is its potential to promote professional growth. I found myself much more confident and persuasive in the recommendations I made in my own field and in other education settings because I had successfully adapted them to the engineering and STEM worlds. I expanded my expertise by working with engineering colleagues, and I hope that they benefitted in the same way from their collaborations with me.

15.4 Challenges

The first challenge I faced was that the communication styles of the engineering colleagues and workshop participants I dealt with differed considerably from those of the educators I was used to dealing with. For example, I learned quickly that being too invitational and giving lots of alternative choices were not what most of the engineers attending our workshops wanted to hear. While they certainly wanted me to suggest teaching strategies and to empower them to make choices suited to their content and students, they wanted the suggestions to be presented in an authoritative style so that they could have confidence in what I was recommending. That took a shift in my thinking and revived my impostor syndrome for a while as I gained confidence in this different style.

I also had to learn a different vocabulary to use in the workshop setting and in multidisciplinary team sessions. I called on colleagues to teach me some basic engineering concepts and tools (flowcharts, free body diagrams, etc.) that I could incorporate into my presentations to increase my credibility with my audience. In one memorable airline flight, Rich taught me the basics of chemical engineering on the back of a napkin: "Stuff comes in ... you do something to it ... more valuable stuff comes out." That seemed a bit oversimplified to me, but it helped! I eventually achieved what I considered a major victory when I was approached after a workshop by a participant who asked me which branch of engineering I was from!

Another challenge I faced that is faced by many—perhaps most—women in engineering was being discounted. I was once told that my participation in an engineering faculty development project was "window dressing" and "nepotism" since Rich and I are married. Putting modesty aside, that view was quickly abandoned once the professor who made the comments worked directly with me in meetings and presentations. I also noticed that many male colleagues would direct all questions to Rich and ignore my presence in meals before our workshops. That pattern also mostly disappeared after we presented, and they saw that I had useful ideas to offer them and valuable responses to their questions, although some discounting still persists.

It is tough to work exclusively outside your own field of expertise. I found that it was important for me to maintain activities in my own field of education. In the late 1990s, a nonengineering colleague—a program evaluator—encouraged me to look into program evaluation, and I took courses leading to certification in that area. She and I have worked together on projects involving program evaluation and qualitative educational research in K-12 and STEM. Those activities have helped me maintain balance and nurture my growth in my discipline.

15.5 Suggestions

If you are a woman working in engineering and your field is in another discipline, here are a few suggestions that may help smooth the way:

- *Choose your projects carefully to be sure you are working with engineering faculty who understand what you will bring to the project and are professionals whom you respect.* Mutual respect is always important for colleagues, but it is critical when working across disciplinary and gender lines.
- *If you are planning to give workshops or make presentations to engineers, partner with an engineer who can get you in the door and provide content-relevant examples as you establish your own credibility.*
- *Be curious, listen to your engineering colleagues, and ask questions.* You are moving into a different culture, and it is important to understand the strengths, needs, and vocabulary within that culture if you hope to have your contributions accepted.
- *Be prepared for pushback when you first make suggestions or question how engineering research is designed and conducted.* You are probably bringing a set of norms and tools the engineers are likely to be unfamiliar with, so you need to be ready to explain why they need to be used and how they should be interpreted.
- *Maintain contacts and work within your home discipline.* Your work in engineering should strengthen and complement your work in your discipline and help you grow professionally.
- *Educate colleagues in your home discipline about the norms in engineering with regard to presentation venues and publications.* You want your work in engineering to "count" in your professional advancement and not hinder your progress. (This was not a factor for me since I moved into a consulting role, but it is very important particularly for women in tenure-track positions.)

And finally, here are suggestions for engineers looking to work with women in fields outside of engineering:

- *Select colleagues you respect, teach them about engineering, and listen to them.* You have chosen to bring in a person from another field because they bring a different and needed perspective. Make sure that you are ready to help them understand the culture you work in and listen to them when they bring ideas to the table. It takes some time to get to a productive working relationship—be patient.

- *Lay the groundwork for them to be respected by other engineers.* Direct questions to them. When introducing them, talk about what they are bringing to your project or development activity. Always treat them collegially.
- *If needed, talk to senior colleagues and department heads in other disciplines to help them understand how publications and presentations work in engineering.* It is almost certain to be different from the norms in their disciplines. Your colleagues need to have their work "count" to make it worth their while to continue to work on your project.

Reference

Felder, R. M., & Brent, R. (2024). *Teaching and learning STEM: A practical guide* (2nd ed.). John Wiley and Sons.

Conclusion: Agents of Change: Reflections of a Lifetime

Daryl E. Chubin

Introduction: A Gendered Perspective

Any contemporary history of women in engineering (and computer science)—with contemporary defined as during my professional lifetime (1973 PhD)—must originate [with a concession to rampant] discrimination that has retarded women's participation and recognition as well as advances within engineering disciplines writ large. In a nod to my intellectual origins, I prefer to offer a *sociology* of women in engineering that attempts to capture the multiple contexts and interventions that have been, if not transformed, then made transparent through research and the unrivaled candor of first-person accounts. Sevo and Chubin (2010) catalogued such a corpus as "bias literacy."

The history of women in engineering is a litany of stories amplified in the preceding chapters by the voices of generations. Engineering education, in particular, will (gratefully) never be the same. Nevertheless, lessons from disparate experiences require codification to be seen as patterns, trends, and prospects. As one who witnessed and modestly contributed to this march of progress, I here assume a role I have played for a half-century as a professor, federal staffer, policy analyst, and advocate—that of an interpreter of what others have reflected and I have observed. Thus, this chapter offers an admittedly self-referential long view, one from a generation or two removed from that of most contributing authors to this publication. The idiosyncrasy of this account stems as well from my gender and perspective as one *outside* the discipline looking in on engineering. What I behold, with profound respect and gratitude, are *agents of change*.

Twenty years ago, colleagues and I took stock of who was and was *not* participating in engineering. In a pair of papers (May & Chubin, 2003; Chubin et al. 2005), we put racial and ethnic diversity as the foremost issue confronting engineering education and the workforce. Those non-white and non-male were sparse in the classroom and the profession. Underlying this sad reality were decades of dubious

J. P. Carpenter (ed.), *Women in Academia*, Women in Engineering and Science,
https://doi.org/10.1007/978-3-031-97938-5

practices that suppressed the participation of mostly prepared and qualified future engineers. This understanding came before intersectionality (Crenshaw, 1989) had entered the mainstream as an empirical approach and a cultural indictment. Intersectionality has led to greater identification and more robust insights into what, in the science education literature, has been termed "contextual mitigating factors" (Gallard Martinez et al., 2019), enabling and perpetuating gender-based discrimination in academic and professional engineering.

Gender, in my scholarship and practice, has long preoccupied me (Chubin and Malcom, 1996; Chubin et al., 2015; Chubin, 2017; U.S. Congress, 1988). As coauthors, colleagues, and subjects for analysis, women in engineering and science embodied problems to be understood, that is, to be studied to reveal categorical differences and similarities in aspirations and career paths. Analytically, the opposite of isolating a variable is contextualizing it and understanding how that variable situates females and other designated minorities. Add time, place, and circumstance to a finding, and causal links emerge.

Categories for Viewing Stasis and Change

Characterizing academic women in engineering requires a deep dive into the participant's categories and analyses that have been used to document both change and resistance to the status quo. In the following, I consolidate the categories that contributors to this volume have employed to tell *their* stories. My narrative is an amalgam of reviews and an interpretation of the trends and triumphs represented within each.

Women in Engineering Programs and Engineering Education

Being ahead of one's time, pioneers outpace their peers. As pathfinders, they are at the forefront of nascent organizations with a vision of what could be versus what is. Like all leaders, pioneers need to attract followers and build a community. What they propose demands vetting and vigorous discussion. Within the traditional disciplinary divisions of engineering—all more than less dominated by men—women have been the historical exceptions, the original classroom *minoritized* population. Their presence attested to their capability and potential, to negotiate the landscape, which the environment neither valued nor encouraged.

The Women in Engineering ProActive Network (WEPAN), founded in 1990, created an affinity group to grow cultures of inclusion in engineering education and professions. Affinity groups come in many guises. Those honoring data provide a form of accountability—measuring how women are integrated into the communities shaping engineering and science while catalyzing human resources for the benefit of all. As a board member, I observed how a national organization could supply a "hand up" (Fort, 1995) by creating a research repository and cataloguing strategies for local interventions. In lieu of "reinventing the wheel" countless times across the United States, these resources offered coping mechanisms for those trapped in hostile work environments and tools for renovating engineering education to welcome the innovations that women introduced into classroom pedagogy and thinking about career prospects.

With the formation of *departments* of engineering education (my familiarity stemmed from an advisory role with Purdue's department), what was a gendered enterprise became an institutionalized recognition of what and how engineering principles could be taught and by whom. There was a distinct consciousness of engineering education—from the faculty composition to the students recruited to the researchable problems defined as urgent and/or outside the mainstream (e.g., environmental degradation and community impacts). This was a kind of antidote to the core of engineering on campus: mobilizing a critical mass of educators to implement promising practices, demonstrating the virtues of peer collaboration, and extending engineering's problem-solving beyond modeling to real-world consequences. Engineering education was ostensibly to inform and foster action on behalf of family, environment, and global well-being.

Women's Affinity Groups, Especially the National Center for Women and Information Technology and AnitaB.org

Affinity compels all of us to gravitate toward those who look like us. There is comfort in similarity. This is the foundational appeal of role models glimpsed at a distance and oftentimes mentors whom we adopt as in-person guides and confidantes. Thus, there should be no surprise that affinity *groups* form and flourish in settings where gender is a significant distinguishing characteristic. Engineering could be Exhibit A for this, hardly the province for nurturing aspiring women students and faculty. Affinity is a compensatory mechanism, the provision of something missing in the broader environment. Testimonials and research documenting dimensions of bias and discrimination provided the transparency lacking in the literature, and the social processes preceding it laid bare this reality.

The first Grace Hopper Celebration of Women in Computing, held in June 1994, featured speeches about computing by prominent women in industry, academia, and government. This milestone evolved with the Institute for Women and Technology (since 2003, AnitaB.org), founded in 1997 on the Xerox PARC campus, and the National Center for Women & Information Technology (NCWIT), chartered in 2004.

Anita Borg's dream was to have comparable numbers of women and men in computer and information science and engineering by 2010, a project known as "50-50 by 2010." Then and now, the fruits of affinity have systematized what is known and still unbidden about the value of women as participants in computer science. The work connects organizations across national boundaries and economic sectors, not just academe, to advance women in technology and K-12 to workforce.

The institutional partners of these clearinghouse organizations comprise a web of expertise, innovation, and opportunity.

Such organizations legitimize gender as a priority—a research focus, a workplace imperative, and a policy marker. The now-defunct Washington, DC-based Commission on Professionals in Science and Technology was a forerunner of these research and advocacy nonprofits. It was founded by Betty Vetter in 1963, who served as editor of, among others, the periodical digests "Scientific, Engineering, Technical Manpower Comments"; "Professional Women and Minorities: A Human Resource Data Compendium"; and "What's Holding Up the Glass Ceiling?" By the turn of the century, CPST was affiliated with AAAS (and supported by the Alfred P. Sloan Foundation) as a low-budget, inside-the-Beltway go-to source for the latest disaggregated data on the production and employment of engineers and scientists. From my perspective as a board member, CPST was indispensable for hosting seminars featuring federal agency and professional society representatives who mined data for policy purposes (Vetter Papers, n.d.). I learned from my participation that statistics do not tell *the* story; they tell many stories, especially about "pipelines" (the parlance back then) inhibiting the flow of women into the technical workforce (Vetter, 1994).

Women at the National Science Foundation and the ADVANCE Program

How does one calibrate the National Science Foundation's (NSF) role in leveling the proverbial playing field for women in science, technology, engineering, and mathematics (STEM)? The leveling has featured the elevation of icons to the directorship of the agency (microbiologist Rita Colwell 1988–2004, astrophysicist France Córdova 2014–2020). But the growth of women in the academic engineering and computing workforce peaked at around 20% in 2021 (ASEE, 2022). Perhaps most telling is the range of initiatives in every NSF directorate dedicated to the advancement of women faculty (from the signature ADVANCE program to more gender-neutral fellowship, center, and supplementary programs that complement individual investigator grants).

When I joined the Education Directorate (EHR, now STEM Education) in 1993, the sex ratio of the professional staff was virtually at parity, leaps and bounds ahead of the six research directorates. There was a critical mass of designated minority women, too. The tenor of meetings differed from that I attended elsewhere in the agency. The mantra was "change the faces around the table of review panelists, and you will change outcomes"—those who become grant awardees—all in the name of rigor and merit. There was a consciousness that EHR promoted and practiced gender and racial equity in its programmatic activities. This "systemic reform" era (Chubin, 1997) focused on pre-K-20 education through collaboration among school districts, cities, and states and formal program evaluation of those complex efforts.

This spirit permeated the ADVANCE program, established in 2001 (with an initial core emphasis, unwittingly or not, on white women). An intervening event, however, set a course for NSF's focus on women's academic careers. In December

1995, co-organizers Sue Rosser and I envisioned NSF *hosting* a Foundation-wide Women & Science Conference for the first time. As the conference report (NSF, 1997) states:

> The purpose . . . was to take stock of the achievements that women have made, assess what works best in the classroom and workplace, and chart a new course to address the challenges that remain.

The conference was a joint effort of the seven NSF directorates. Seven hundred women and men from colleges and universities, industry, nonprofit groups, schools, and community groups assembled in Washington, DC, . . . to take part in plenary sessions, breakout sessions, and a special showcase of posters and demonstrations. Breakout sessions were organized by scientific and engineering disciplines and by cross-cutting themes: research-education infrastructure, impact of technology, family issues, shattering preconceptions, bridging education and workforce transitions, and changing curriculum and instruction.

Prior to this conference, EHR's Programs for Women and Girls championed such issues. But they were hardly an NSF priority. Two questions guided the conference:

- In what creative ways could the National Science Foundation increase the participation of and opportunities for women in science, mathematics, engineering, and technology?
- How could this conference be used to stimulate change in local campuses and in communities?

Arguably, ADVANCE was the Foundation's response—to this trailblazing conference and doubtless many other events—as well as a public commitment to take gender equity "to scale." The rest, as they say, is history.

I subsequently served on a half dozen ADVANCE external advisory committees from the first cohort onward and learned first-hand of the barriers that women STEM faculty had confronted. Interviews with ADVANCE participants, from faculty to administrators, revealed a wide range of experiences and limiting situations (Freire, 2005), many of which had been structurally enshrined in a male-dominated cultural landscape. Measuring disparities in status, salary, workload, and trappings of success had just begun. Today, data convey but still do not ensure accountability. Templates of climate surveys abound. Research repositories harbor tools, and webinars dispense strategies. ADVANCE is a testament to shared intelligence—and transparency of the wrongs to be righted. Sexual harassment, microaggressions, and assorted biases remain a contemporary scourge, not an antiquated anomaly.

Engineering was perhaps no worse than other disciplines on the continuum of misdeeds. But the number of women in engineering fields was decidedly smaller. Departments with a single female faculty member are hard to miss. Colleges of engineering with hundreds of faculty and less than 10% women are noticeable. Students notice, too, which compounds the problem of recruitment and advancement (Margolis & Fisher, 2002). NSF programs address such inequities or at least provide resources that legitimize efforts on campus—the NSF imprimatur of merit and quality—to reduce gender biases and diversify the professoriate. Money talks, but local leaders must listen—and act to implement institutional change.

Women Engineering Academic Leaders: Deans, Provosts, Presidents

The Society of Women Engineers reminds us that in 1984, electrical engineer Eleanor Baum became the first woman appointed Dean at Pratt Institute of an engineering program in the United States. Today, the American Society for Engineering Education counts more than 80 women leading engineering colleges, schools, or programs in the United States out of more than 300 programs (SWE, 2023). This is an unmistakable progress. But what a long, hard road it has been.

According to NSF data, about one in five undergraduate engineering students nationally is a woman. Baccalaureate degrees in mathematics and computer science leveled by 2008 at one in four. Yet doctoral degrees awarded to women in engineering increased to only 21% over four decades (NCSES, 2023).

The higher up the leadership chain in institutions of higher education, the scarcer women become. Statistically, women engineers who ascend to university presidencies represent few links in the chain. A Women's Power Gap report found that only 30% of presidents at US research universities are women (Silbert, 2023). Those who have excelled since the 1970s are familiar names—Mildred Dresselhaus and Sheila Widnall come to mind—who symbolically paved the way for those such as chemical engineer Gilda Barabino, the first woman and first Black President of Olin College of Engineering. She defines leadership this way:

> I believe it's very important for leaders to be bold, courageous, unapologetic, and equity minded. If you want to be serious about your mission, you must be willing to step out there and provide that kind of leadership … and be adaptable. What might have worked at one point in time, may not work today. The world is rapidly changing, and leaders must be agile and prepared to meet these changes, as well as the needs of the students that we are bringing in now (Emery, 2022).

Notably, Barabino was one of the signatories of an all-women group of university Presidents and Deans of Engineering from six of America's leading research universities who issued an Open Letter in April 2023 supporting the CHIPS and Science Act. The letter states, "Building a more heterogeneous, job-ready labor force demands that higher education, private industry, and the federal government coalesce and act at an unprecedented level and pace" (All Women Group, 2023). The letter also states that such forward-looking and engaged leadership should be hailed for citing societal impact, creation of educational on-ramps, investment in engineering education infrastructure, and permeability between higher education and industry. This is the hallmark of women leaders in academe.

Women National Professional Society Leaders, Including the National Academies

If the pinnacle of academic leadership is a university presidency, then both a stepping-stone and unflagging ally are the professional societies that boost visibility and support in myriad ways the mission of gender equity. In addition to those non-profits noted above, disciplinary organizations such as the Computing Research Association and cross-cutting ones such as the American Association of University Women abound. All are bastions of inter-generational mentoring, machines of net-working opportunity, and trusted disseminators of research-based resources. (They, too, have cultivated vaunted leaders like the editor of this volume, a dynamic dean at Campbell University, and former President of the American Society for Engineering Education.)

Some of these gender-centered nonprofits emphasize information technology or imparting coding skills to girls as instruments of broadening participation in engineering and computing both within and outside academic settings. Notably, launched in 2001, the EngineerGirl website (engineergirl.org) is designed to bring national attention to the myriad opportunities that engineering represents for girls and women. It lists 37 professional societies—from the stalwart IEEE Women in Engineering to the decade-old Black Girls Code—dedicated to attracting, informing, and supporting women at various junctures of their engineering-related careers.

In 2001, I moved from NSF to the then-largest minority-serving undergraduate scholarship organization, the National Action Council for Minorities in Engineering (NACME, https://www.nacme.org/about-us). NACME, too, was founded by the National Academy of Engineering (NAE), in 1974. A generation later, EngineerGirl appeared, with this rationale:

> Diversity of thought is crucial to creativity, and by leaving women out of the process of innovation we lose a key component of diversity and stifle innovation. We want the creative problem-solvers of tomorrow to fully represent the world's population, because they will be the ones to ensure our health, happiness, and safety in years to come (https://www.engineergirl.org/10851/About).

Pivotal are resources provided by professional societies that promote persistence and reduce isolation by building community and social capital (Campbell-Montalvo et al., 2022), hence the emphasis on belonging, networking, and mentoring to combat the unwelcoming environments women engineers often encounter.

The National Academies of Science, Engineering, and Medicine indeed loom over the constellation of colleges, universities, nonprofits, and corporations that employ, compete for, and recognize women in STEM (NASEM, 2022). Recently, NASEM announced a new Action Collaborative on Transforming Trajectories for Women of Color in Tech that comprised of 35 higher education, national laboratory, and governmental institutions (https://www.nationalacademies.org/our-work/action-collaborative-on-transforming-trajectories-for-women-of-color-in-tech). But the National Academies' election of women has long lagged—currently constituting around 20% (NAE, 2023) of its total membership. While hailed as the pinnacle of achievement, membership in the National Academies still reflects the wider, entrenched US academic research culture.

Academic Women, Positionality, and Intersectionality

As demonstrated above, one measure of progress in charting STEM student and workforce demographics is the *disaggregation* of data that discloses differences at a granular level: not just diversity, but also inclusive excellence, and not only broadening participation, but also productive leadership. The era of merely counting inputs gives way to dashboard monitoring and transparency of outcomes that are essential in making explicit socioculturally embedded limit situations (Gallard Martinez et al., 2019).

Ironically, this same era is defined by a Supreme Court (Students for Fair Admissions, 2023) decision forbidding the use of affirmative action in college admissions and, stunningly, an avalanche of state prohibitions of anything smacking of DEI (diversity, equity, and inclusion). Programs, positions, and language are all being eradicated (and redefined) on campuses across the United States. The overturning of *Grutter* has declared open season on preferences that favor one group over another in competitions for scarce educational resources—access, financial aid, student services, etc. Women, a numerical majority, are among those who retain a disadvantaged status in allocating what matters.

Such cultural backlash is clarified through two analytical lenses, positionality, and intersectionality, which can be deployed to help dissect how women in engineering have fared over time. Each is implied by what has been observed heretofore in this chapter. Now, they are made explicit.

Arguably controversial, "positionality statements" have been adopted by many journals, scholars, and nonprofit practitioners. The function of declaring positionality is to link identity to action. Positionality declares what analysts seldom disclose: how they think is conditioned by who they are. This, in turn, demonstrates how attributes of authors/researchers influence their methods and choice of problem for study. Additionally, positionality can also make transparent how people are situated in a status quo that defines people based on their gender and race, as examples. Such clarity frames knowledge as a product of a point of view and approach (Zamzow, 2023). It peels away the veneer of researcher-independent objectivity and presents a more candid and fallible interpretation of findings and outcomes.

The implication is that humans habitually attribute positions based on social characteristics. We stereotype and *assign* positionality all the time, leading to "doxas" (common beliefs or personal opinions). To wit:

Women can't be engineers.

Black girls can't code

Asians can't lead

White men can't understand

Enter the "scholar activists" who position themselves as bridges between the academy and the minoritized communities—defined by race, ethnicity, class, and identity—they seek to serve (Cypress, 2023). Scholar activism is an expression of positionality that asserts the inevitable influence of experiences, interests, and biases on those who conduct research. Therefore, it is futile to adopt the pretense of neutrality. In contrast, traditional scholars view the pursuit of knowledge as intrinsically valuable while trying to correct for biases. Advocacy is a form of activism, and scholar-activists lean into biases as an exploratory and, sometimes, an explanatory tool.

The sociology of knowledge posited all this more than a century ago (Berger & Luckmann, 1966). In the current context, consider the identifications now employed to characterize undergraduate students: first generation, Pell-eligible, special needs, English-speaking as a second language, non-Christian, and LGBTQ+, among others. Thus, one might ask whether a study of women undergraduates in computer science conducted by a female researcher yields different findings from a study conducted by a male, an African-American, or any of the identifications listed above. But positionality should not be mistaken for naked identity politics. Nor should scholar activism. Rather, positionality uses identity as a social construct to explain, elaborate, and clarify reality from a perspective of marginality or underrepresentation.

For what it is worth, this view does not mean that all women academic scholars are or should be activists. Yet, a proclivity for activism might highlight what remains obscure in traditional analyses. I offer this as a White cisgender man who identifies with two of the attributes noted above—and hope I am not guilty of "mansplaining."

Intersectionality suggests a more specific focus on marginalizing race and class as discriminatory and political acts (Crenshaw, 1989). It further amplifies the "lived experiences" reflected by positionality as shaping the hierarchies of the social order—favoring some and undervaluing (even oppressing or disenfranchising) others. The application of intersectionality to engineering and science beckons us to dive beneath the crosstabs of participation and examine the conditions that leave women—and particularly women of color—behind in achieving positions of leadership and influence as faculty, deans, provosts, and university presidents (Ong et al., 2020).

The deeper dive afforded by intersectionality shows that debt burden and time to degree are greater for STEM PhDs who are Black (Weissman, 2023). By inference, women of color compared to men lag even further behind. But their vanishingly small number precludes reporting that risks violating their anonymity. That phenomenon is salient data. It points to the vagaries of mentoring and the hazards of harassment that compound the challenges these women face. Traditional NSF statistics revealed neither such complexities nor disparities. It took researchers' self-reflective consciousness to probe the correlates of underrepresentation and expose the hardship-strewn pathways taken (e.g., Charleston et al., 2014).

Such statistical revelations are also a reminder that depending on the interpersonal setting, *anyone* can be minoritized or made to feel as though they are designated a minority in a way that is harmful and unfair. How one is seen (or not) assigns them a status: less than, different from, investable, and dispensable. The label, while not all-defining, is stigmatizing.

Gender and race thus appear as key antecedents of career progression—individually and organizationally. Together, their display of opportunity gained and denied attests to the analytical power of positionality and intersectionality. The lesson is clear: no study of careers should rely on demographics alone. Culture and context matter in the amassing of knowledge. The biases inherent in us all demand sophistication of theory and method in constructing a fuller, more transparent picture of attribution and achievement of women in STEM (Malcom & Chubin, 2005).

Rhetoric and Research

Connecting various studies, testimonials, and advocacy reports produces a mélange of impressions with gender at the intersections: policy breeds programs through funding awarded to organizations led by accomplished individuals. Viewed separately, each component is magnified in significance; taken together, a complex of relationships is visible and celebrated. Title IX created opportunities for federal agencies such as NSF to formulate the ADVANCE program and distribute resources to universities across the United States. Leaders of and participants in those programs fostered a community and network for dissemination, experimentation, and adaptations to support, legitimize, and promote women in engineering and science roles (Rosser et al., 2019). What was once compensatory action has become more structural and routine.

Yet, sclerotic cultures retard organizational change. Despite the exhilaration of individual stories of women engineers excelling and overcoming, national trend data are more sobering. Among the latter are those who caution that advocates' rhetoric, spurred by cherry-picked data and media hyperbole, often outstrips the research record. Williams and Ceci (2023) summarize:

> In journal articles, we discuss claims of sexism in tenure-track science positions that have appeared in the most prestigious scientific outlets, as well as in statements by top professional societies and blue-ribbon reports... These include bias against women in six key domains: tenure-track hiring, journal peer reviewing, grant funding, letters of recommendation for faculty applicants, salary, and teaching evaluations.
>
> Nowhere have we asserted that the academy is devoid of sexism. The entire structure of the tenure probationary period and evaluation process is harder on women than men due to women's far-tighter reproductive window. This could and should be changed. .. Salaries should be aggressively reviewed and gender gaps eliminated—all that is needed is an administration's will.

Even the most enlightened administrator, however, cannot decree workplace equity. Rather, such leaders must act to build a culture that applies even-handedly the criteria for promotion, tenure, and remuneration.

Williams and Ceci's findings on gender-based discrimination were recently reinforced by other data on the persistence of women's lagging pay equity (Moody, 2023). This is reminiscent of "A Study on the Status of Women in Science at MIT" (1999, since revisited in Zernike, 2023). Though validated, this landmark study also warned of backsliding on fairness metrics once progress is made. Too often, it seems, the greatest obstacle women face on the path to leadership is a lack of trust and transparency—what is termed in the corporate world the "broken rung":

> Conventional wisdom says that women hit a "glass ceiling" as they advance that prevents them from reaching senior leadership positions. In reality, the biggest obstacle that women face is the first step up to manager … (Leanin.org, 2019).

The academic counterpart to this hierarchy would be the leadership ladder from department chair to college dean to institution-wide position. The relative dearth of such ascensions in academic engineering and computer science speaks to an ingrained institutionalized sexism. In the MIT case above, President Charles Vest, a mechanical engineer, is credited for welcoming the study, acknowledging the existence of the discrimination found, and taking steps to redress it. Today, we look to a woman at the top to accelerate the pace of change in policy and practice.

"Fix the system, not the woman" is a well-known battle cry of those who study gender disparities in STEM (Meiksins & Layne, 2023). But it takes agents of change to disrupt and realign the system to be more just and open. A vital aspect of such agency is the gatekeeping function exercised by leaders located in key positions. Gatekeepers are active filters of quality and opportunity. Just as some information, but not all, gets passed along, so too are some system participants selected for confronting new challenges and the prospect of making contributions to the organization they inhabit. Overlooking them due to attributes unrelated to their performance and talent penalizes not only the candidate, but also the organization. This makes gatekeepers—a combination of evaluator, mentor, and advocate—a breed of talent scout.

Male domination of complex organizations—universities and corporations alike—has no doubt squandered talent (much of it female). Leaders who recognize the ongoing tendency to pass over women faculty and administrators are positioned to change the roster. The question is whether they work to do so or not.

Whither Agents of Change?

This chapter began with an expressed preference for a sociology of women in engineering and science. Three decades ago, a sociologist established one of the first curricular programs to study women, science, and technology. Through her scholarship and national leadership, Mary Frank Fox (e.g., Fox, 2001; Fox, 2020) has helped to pioneer gender equity as a national imperative. Her prescription for change (Fox, 2020) is instructive:

> . . . clear, written guidelines can help create a fair chain of command and responsibility. Committees, chairs of committees, and deans working within this system are all needed to ensure a level playing field.

In Fox's own words, "There is nothing impossible about this. It takes organizational will and it takes attention and responsibility. But it is not magic" (Walton, 2018). The persistent lack of gender diversity seems to follow a path of least resistance, not malevolence. Yet, cultural atavisms dominate: inequality is a central feature of science, and gender relations in academe remain hierarchical. Thus, we talk in grandiose terms of needed "institutional transformation." The problem is that such transformation, like "scientific revolution," is recognized only in retrospect.

In the here and now, agents of change must resort to a variety of tactics to move the culture of bias. Scholarship is one tactic at a high premium in higher education. Funding success is another. Scholar activism is a third. But the one demonstrably effective, hard-won tool of transformation is *leadership*. How does one rise to positions of authority, gatekeeping, resource allocation, and opinion leadership? (A recent exemplar who jumps to mind is Shirley Ann Jackson of RPI.) Whether cracking glass ceilings or scaling rungs on career ladders, women who ascend to leadership *advocate*—raising the sisters, brothers, and *institutions* "en route."

In this sense, women in engineering and science are no different from women leaders in business, government, athletics, and the arts. All must negotiate complex organizations in an era of technological disruptions. Yet, it is the *social* disruptions that focus attention on the implicit biases and explicit inequities that continue to sap institutions of their creativity and productive potential. Women are breaking these structural barriers, but at a high cost to them personally. Meritocracy is kind neither to the biological clock nor to the mental strains modern work imposes. "Imposter syndrome" may be clinically gender neutral—we have all felt doubts upon assuming a new role on a new rung—but women (and ethnic minorities) in disciplines that value "brilliance" manifest the affliction far more than men (Clance & Imes, 1978; Ibrahim, 2021).

Perhaps part of deconstructing the contemporary workplace is to calibrate concepts like merit, imposter, and brilliance, which are more contextually pliable than fixed. Perhaps achievement should be seen, despite the permeation of women in late twentieth- and twenty-first-century academic engineering and computer science, as a process of creating inroads and pathways through social order, invoking policy, and effecting accountability so as to innovate the conditions that facilitate careers. Above all, we must be mindful that promoting leaders is not enough; enabling them to succeed is the organization's reciprocal responsibility.

In a concluding example, the venerable *New York Times* (Moreno, 2023) was recently taken to task for not crediting women researchers in the development of artificial intelligence technologies:

> Shame on The New York Times for omitting every single woman who has shaped A.I. Not a single woman made this list. Not even Stanford Professor Fei-Fei Li, who founded AI4All and launched Stanford's Human-Centered A.I. Institute, was mentioned. The bias and prejudice are so obvious. Intentional erasure is a form of violence and misogyny (https://lnkd.in/gh-M3C5Q).

Slights and oversights like this have consequences for individual careers and, moreover, for the ecosystem shaping public understanding of the creative process. Such misrepresentations are a form of gatekeeping in which fresh ideas and certain faces are kept out (e.g., O'Neil, 2023). They need to be *called out* as agents of change—women and men alike—continue to operate in these hostile-to-indifferent environments. Their conduct must highlight and simultaneously strive to *reduce* the significance of gender in the coming history of engineering and science.

References

All Women Group of University Presidents, Deans of Engineering from Six of America's Leading Research Universities Issue an Open Letter Supporting the CHIPS and Science Act. (2023). Retrieved from https://www.prnewswire.com/news-releases/all-women-group-of-university-presidents-deans-of-engineering-from-six-of-americas-leading-research-universities-issue-an-open-letter-supporting-the-chips-and-science-act-301803138.html

ASEE. (2022). *Engineering and engineering technology by the numbers* (2021st ed.) Retrieved from https://ira.asee.org/wp-content/uploads/2022/09/Engineering-and-Engineering-Technology-by-the-Numbers-2021.pdf

Berger, P., & Luckmann, T. (1966). *The social construction of reality: A treatise in the sociology of knowledge*. Doubleday.

Campbell-Montalvo, R., Kersaint, G., Smith, C. A., Puccia, E., Sidorova, O., Cooke, H., Wao, H., Martin, J. P., Skvoretz, J., MacDonald, G., & Lee, R. (2022). The influence of professional engineering organizations on women and underrepresented minority students' fit. *Frontiers in Education, 6*, 755471. https://doi.org/10.3389/feduc.2021.755471

Charleston, L., Adserias, R., Lang, N., & Jackson, J. (2014). Intersectionality and STEM: The role of race and gender in the academic pursuits of African American women in STEM. *Journal of Progressive Policy and Practice, 2*(3), 273–293.

Chubin, D. (1997). Systemic evaluation and evidence of education reform. In D. Bartels & J. Sandler (Eds.), *Implementing science education reform: Are we making an impact?* American Association for the Advancement of Science.

Chubin, D., Didion, C., & Beoku-Betts, J. (2015). Promising programs—a cross-national exploration of women in science, education to workforce. In W. Pearson Jr., L. M. Frehill, & C. L. McNeely (Eds.), *Advancing women in science: An international perspective* (pp. 275–292). Springer.

Chubin, D. E. (2017). Practices in engineering education (as viewed through four frames). In *Transforming Undergraduate Education in Engineering Phase III: Voices on Women's Participation and Retention. Workshop Report* (pp. 61–72). American Society for Engineering Education. Retrieved December 6, 2023, from http://docs.asee.org/public/TUEE/PhaseIII/Artifacts/TUEEPHASEIIIReport.pdf

Chubin, D. E., & Malcom, S. M. (1996). Policies to promote women in science. In C.-S. Davis et al. (Eds.), *The equity equation: fostering the advancement of women in the sciences, mathematics, and engineering* (pp. 1–28). Jossey-Bass.

Chubin, D. E., May, G. S., & Babco, E. L. (2005). Diversifying the engineering workforce. *Journal of Engineering Education, 94*(1), 73–86. Retrieved December 6, 2023, from https://doi.org/10.1002/j.2168-9830.2005.tb00830.x

J. P. Carpenter (ed.), *Women in Academia*, Women in Engineering and Science,
https://doi.org/10.1007/978-3-031-97938-5

Clance, P. R., & Imes, S. (1978). The imposter phenomenon in high achieving women: dynamics and therapeutic intervention. *Psychotherapy: Theory, Research & Practice, 15*(3), 1–8.

Crenshaw, K. (1989). Demarginalizing the intersection of race and sex: A black feminist critique of antidiscrimination doctrine, feminist theory and antiracist politics. *University of Chicago Legal Forum (1989, 1)*. Retrieved December 12, 2023, from http://chicagounbound.uchicago.edu/uclf/vol1989/iss1/8

Cypress, R. (December 13, 2023). The abuse of scholar activism. *Inside Higher Ed*. Retrieved December 18, 2023, from https://www.insidehighered.com/opinion/views/2023/12/13/scholar-activism-contributes-antisemitism-opinion

Emery, M. (2022). Mission accomplished: Gilda barabino shares insights on leadership. Retrieved December 13, 2023, from https://uh.edu/provost/news/news-releases/accomplished-leaders-recap-2022.php

Fort, D.C. (Ed.) (1995). A hand up: women mentoring women in science paperback.

Fox, M. F. (2001). Women, science, and academia: Graduate education and careers. *Gender and Society, 15*, 654–666, Retrieved December 18, 2023, from. https://doi.org/10.1177/089124301015005002

Fox, M. F. (2020). Gender, science, and academic rank: key issues and approaches. *Quantitative Science Studies, 1*, 1001–1006. https://doi.org/10.1162/qss_a_00057

Freire, P. (2005). *Education for critical consciousness* (2nd ed.). Continuum.

Gallard Martinez, A. J., et al. (2019). Discerning contextual complexities in STEM career pathways: insights from successful Latinas. *Cultural Studies of Science Education, 14*, 1079–1103. https://doi.org/10.1007/s11422-018-9900-2

Ibrahim, M. (August 13, 2021). Women feel like imposters in disciplines that value 'brilliance'. *Science*. Retrieved December 18, 2023, from https://www.science.org/content/article/women-feel-imposters-disciplines-value-brilliance

leanin.org. (2019). The 'broken rung' is the biggest obstacle women face. Retrieved December 8, 2023, from https://leanin.org/women-in-the-workplace-2019#

Malcom, S. M., & Chubin, D. E. (2005). Letting nurture take its course. In D. Fort (Ed.), *A hand up: Women mentoring women in science* (pp. 249–266). Association for Women in Science.

Margolis, J., & Fisher, A. (2002). *Unlocking the clubhouse: Women in computing*. MIT Press.

May, G., & Chubin, D. (2003). A retrospective on undergraduate engineering success for underrepresented minority students. *Journal of Engineering Education, 92*, 27–40.

Meiksins, P., & Layne, P. (2023). Women in engineering: Analyzing 20 years of social science literature. *SWE: Magazine of the Society of Women Engineers*, vol. 69, Conference 2023. Retrieved December 8, 2023, from https://magazine.swe.org/lit-review-22/

MIT Faculty Newsletter. (1999). A study on the status of women in science at MIT, vol. XI, No. 4.. Retrieved December 16, 2023, from http://web.mit.edu/fnl/Women/Women.Html

Moody, J. (December 6, 2023). Survey: Pay equity lags for higher Ed administrators. *Inside Higher Ed*. Retrieved December 16, 2023, from https://www.insidehighered.com/news/quick-takes/2023/12/06/survey-pay-equity-lags-higher-ed-administrators

Moreno, J. E. (2023). *Who's who behind the dawn of the modern artificial intelligence movement?* The New York Times. Retrieved December 31, 2023, from https://www.nytimes.com/2023/12/03/technology/ai-key-figures.html?unlocked_article_code=1.KU0.pwPM.TFIRQXccIfi-&smid=em-share

NAE (National Academy of Engineering). (2023). National Academy of Engineering Elects 106 Members and 18 International Members. Retrieved December 23, 2023, from https://www.nae.edu/289843/NAENewClass2023

NASEM (National Academies of Sciences, Engineering, and Medicine). (2022). *Transforming trajectories for women of color in tech*. The National Academies Press. Retrieved December 23, 2023, from. https://doi.org/10.17226/26345

NCSES (National Center for Science and Engineering Statistics). (2023). *Diversity and STEM: Women, minorities, and persons with disabilities*. National Science Foundation.

Retrieved December 30, 2023, from https://ncses.nsf.gov/pubs/nsf23315/report/science-and-engineering-degrees-earned

NSF (1997). *Women & Science: celebrating achievements, charting challenges, national science foundation conference report* (December 1995), http://www.ehr.nsf.gov/conferences/women95.htm

O'Neil, L. (August 12, 2023). These women tried to warn us about A.I. rolling stone. Retrieved January 3, 2024, from https://www.rollingstone.com/culture/culture-features/women-warnings-ai-danger-risk-before-chatgpt-1234804367/

Ong, M., Jaumot-Pascual, N., & Ko, L. T. (2020). Research literature on women of color in undergraduate engineering education: A systematic thematic synthesis. *Journal of Engineering Education, 109*(3), 347–615. Retrieved December 30, 2023, from. https://doi.org/10.1002/jee.20345

Rosser, S., Barnard, S., Carnes, M., & Munir, F. (2019). Athena SWAN and ADVANCE: Effectiveness and lessons learned. *Viewpoints, 393*(10171), 604–608. Retrieved December 12, 2023, from. https://doi.org/10.1016/S0140-6736(18)33213-6

Sevo, R., & Chubin, D. (January, 2010). *Bias literacy: A review of concepts in research on gender discrimination and the U.S. context.* Retrieved November 20, 2023, from.. https://doi.org/10.4018/978-1-61520-657-5.ch002

Silbert, A. (November, 2023). *Transparency is first step toward more diversity in leadership* (Inside Higher Ed.) Retrieved November 21, 2023, from https://www.insidehighered.com/opinion/views/2023/11/20/more-data-needed-ensure-diversity-leadership-opinion

Students for Fair Admissions, Inc. (SFFA) v. President & Fellows of Harvard College (Harvard) and SFFA v. University of North Carolina (UNC), Nos. 20-1199 & 21-707 (2023). Retrieved January 8, 2024, from https://www.scotusblog.com/case-files/cases/students-for-fair-admissions-inc-v-president-fellows-of-harvard-college/

SWE (Society of Women Engineers). (2023). State of Women in Engineering 2021: Women Deans and Directors of Engineering Programs in the U.S. *Magazine of the Society of Women Engineers, Conference 2023*, vol. 69, Retrieved December 16, 2023, from https://magazine.swe.org/women-deans-and-directors-of-engineering-programs-in-the-u-s/

U.S. Congress, Office of Technology Assessment. (1988). *Educating scientists and engineers—Grade school to grad school.* U.S. Government Printing Office.

Vetter, B. M. (1994). The next generation of scientists and engineers: Who's in the pipeline? In W. Pearson Jr. & A. Fechter (Eds.), *Who will do science? Educating the next generation* (pp. 1–19). Johns Hopkins University Press.

Vetter Papers, B. M., & MS 582. (n.d.). *Archives of women in science and technology.* Iowa State University, Special Collections. Retrieved December 11, 2023, from https://findingaids.lib.iastate.edu/spcl/manuscripts/MS582.pdf

Walton, M. (May 11, 2018). Science, technology, and gender: a sociologist's quest for equality. Retrieved December 16, 2023, from https://www.aaas.org/science-technology-and-gender-sociologists-quest-equality

Weissman, S. (November 27, 2023). Black scientists with STEM PhDs face deep disparities. *Inside Higher Ed.* Retrieved December 6, 2023, from https://www.insidehighered.com/news/diversity/race-ethnicity/2023/11/27/new-report-finds-disparities-black-stem-phds

Williams, W. M., & Ceci, S. J. (November 30, 2023). How sexist is science? *The Chronicle of Higher Education.* Retrieved December 6, 2023, from https://www.chronicle.com/article/how-sexist-is-science

Zamzow, R. (2023). Should scientists include their race, gender, or other personal details in papers? *Science, 382*(6670), 501. Retrieved December 16, 2023, from https://www.science.org/content/article/should-scientists-include-their-race-gender-or-other-personal-details-papers

Zernike, K. (2023). *The exceptions: Nancy hopkins, MIT, and the fight for women in science.* Scribner.

Index

J. P. Carpenter (ed.), *Women in Academia*, Women in Engineering and Science, https://doi.org/10.1007/978-3-031-97938-5

MIX
Papier aus verantwortungsvollen Quellen
Paper from responsible sources
FSC® C105338

If you have any concerns about our products,
you can contact us on
ProductSafety@springernature.com

In case Publisher is established outside the EU,
the EU authorized representative is:
Springer Nature Customer Service Center GmbH
Europaplatz 3, 69115 Heidelberg, Germany

Printed by Libri Plureos GmbH
in Hamburg, Germany